Recent Advancements in Connected Autonomous Vehicle Technologies

Volume 3

Series Editors

Yanjun Huang, School of Automotive Studies, Tongji University, Shanghai, China

Wei He, University of Science & Technology, Beijing, China

Xiaosong Hu, Automotive Engineering, Chongqing University, Chongqing, China

Shengbo Eben Li, School of Vehicle and Mobility, Tsinghua University, Beijing, China

Weichao Sun, The Research Institute of Intelligent Control and Systems, Harbin Institute of Technology, Harbin, Heilongjiang, China

Guodong Yin, School of Mechanical Engineering, Southeast University, Nanjing, Jiangsu, China

Hui Zhang, School of Transportation Science and Engineering, Beihang University, Beijing, China

Wanzhong Zhao, College of Energy and Power Engineering, Nanjing University of Aeronautics and Astronautics, Nanjing, Jiangsu, China

Bing Zhu, College of Automotive Engineering, Jilin University, Changchun, Jilin, China

This Springer book series shows control theory applications and the latest achievements in the field of the autonomous vehicle. It emphasizes the practical application of various control methods, with real-world experimental validations. New and developing control algorithms, including data-driven control, robust control, cooperative control, and predictive control, are discussed and applied in the connected autonomous vehicle to improve energy consumption, vehicle safety, passenger feeling, and transportation efficiency.

The book series aims to introduce essential and recent achievements in the autonomous vehicle to researchers and engineers. The individual book volumes in the series are thematic. The goal of each volume is to give readers a comprehensive overview of how developing control algorithms can be used to improve vehicle performance and transportation efficiency. As a collection, the series provides valuable resources to a broad audience in academia, the engineering research community, industry, and anyone else who are looking to expand their knowledge of the autonomous vehicle.

Yue Cao · Yuanjian Zhang · Chenghong Gu
Editors

Automated and Electric Vehicle: Design, Informatics and Sustainability

Springer

Editors
Yue Cao
School of Cyber Science and Engineering
Wuhan University
Wuhan, Hubei, China

Suzhou Research Institute of Wuhan
University
Suzhou, China

Yuanjian Zhang
Department of Aeronautical
and Automotive Engineering
Loughborough University
Loughborough, Leicestershire, UK

Chenghong Gu
Department of Electronic and Electrical
Engineering
University of Bath
Bath, UK

This work was supported by Suzhou Municipal Key Industrial Technology Innovation Program (SYG202123).

ISSN 2731-0027 ISSN 2731-0035 (electronic)
Recent Advancements in Connected Autonomous Vehicle Technologies
ISBN 978-981-19-5753-6 ISBN 978-981-19-5751-2 (eBook)
https://doi.org/10.1007/978-981-19-5751-2

© The Editor(s) (if applicable) and The Author(s), under exclusive license to Springer Nature Singapore Pte Ltd. 2023
This work is subject to copyright. All rights are solely and exclusively licensed by the Publisher, whether the whole or part of the material is concerned, specifically the rights of translation, reprinting, reuse of illustrations, recitation, broadcasting, reproduction on microfilms or in any other physical way, and transmission or information storage and retrieval, electronic adaptation, computer software, or by similar or dissimilar methodology now known or hereafter developed.
The use of general descriptive names, registered names, trademarks, service marks, etc. in this publication does not imply, even in the absence of a specific statement, that such names are exempt from the relevant protective laws and regulations and therefore free for general use.
The publisher, the authors, and the editors are safe to assume that the advice and information in this book are believed to be true and accurate at the date of publication. Neither the publisher nor the authors or the editors give a warranty, expressed or implied, with respect to the material contained herein or for any errors or omissions that may have been made. The publisher remains neutral with regard to jurisdictional claims in published maps and institutional affiliations.

This Springer imprint is published by the registered company Springer Nature Singapore Pte Ltd.
The registered company address is: 152 Beach Road, #21-01/04 Gateway East, Singapore 189721, Singapore

Contents

1 Energy Efficient Control of Vehicles 1
Yuanjian Zhang and Zhuoran Hou

2 Battery Management System of Electric Vehicle 23
Yuanjian Zhang and Zhuoran Hou

3 Speed Forecasting Methodology and Introduction 45
Yuanjian Zhang and Zhuoran Hou

4 Eco-Driving Behavior of Automated Vehicle 69
Yuanjian Zhang and Zhuoran Hou

5 Service Planning and Operation for Autonomous Valet Parking ... 81
Ziyi Hu, Jianyong Song, Yue Cao, and Yongdong Zhu

6 Navigation Service Optimization for Electric Vehicle 99
Xinyu Li, Yue Cao, and Shuohan Liu

**7 AI-Based GEVs Mobility Estimation and Battery Aging
Quantification Method** 117
Shuangqi Li and Chenghong Gu

**8 Multi-objective Bi-directional V2G Behavior Optimization
and Strategy Deployment** 135
Shuangqi Li and Chenghong Gu

9 Local Energy Trading with EV Flexibility 153
Shuang Cheng, Da Xie, and Chenghong Gu

10 A Review of the Trends in Smart Charging, Vehicle-to-Grid 175
Ridoy Das, Yue Cao, and Yue Wang

**11 Communication and Networking Technologies in Internet
of Vehicles** 197
Yujie Song, Jianyong Song, Sihan Qin, and Yue Cao

12 The Overview of Non-orthogonal Multiple Access in Vehicle-to-Vehicle Communication 219
Lei Wen and Yue Cao

13 Decentralized Trust Management System for VANETs 241
Yu Wang, Jianyong Song, Yu'ang Zhang, and Yue Cao

14 Intrusion Detection System for Connected Automobiles Security 257
Abdul Majid Jamil, Jianhua Zhou, Di Wang, Hassan Jalil Hadi, and Yue Cao

Chapter 1
Energy Efficient Control of Vehicles

Yuanjian Zhang⊙ and Zhuoran Hou

Abstract Electric vehicles (EVs) have the advantages of energy saving and environmental protection, which are favoured by major vehicle companies nowadays. However, the problem of how to effectively improve the economy has been a hot spot and difficult research point of the vehicle control strategy. Therefore, this chapter introduced the mainstream algorithms currently used as energy management strategies, and analysed the advantages of each method. This chapter begins with an introduction to energy integrated control for electric vehicles. Since the control scheme is related to architecture, this chapter then introduces the common architectures of EVs. Finally, the rule-based energy management strategy and the optimization-based energy management strategy are highlighted, and the vehicle architectures to which the different strategies are adapted are analyzed. Finally, the development and characteristics of the strategies are summarized.

Keywords Electric vehicle · Energy management strategy · Energy consumption · Rule-based · Optimization-based

List of Abbreviations

EVs	Electric Vehicles	PHEV	Plug-in Hybrid Electric Vehicle
SOC	State of Charge	DMC	Dynamic Matrix Control
DP	Dynamic Programming	MPHC	Model Predictive Heuristic Control
PMP	Pontriagin Minimum Principle	MPC	Model Predictive Control

Y. Zhang (✉)
Department of Aeronautical and Automotive Engineering, Loughborough University, Leicestershire, UK
e-mail: y.y.zhang@lboro.ac.uk

Z. Hou
State Key Laboratory of Automotive Simulation and Control, Jilin University, Changchun, China

© The Author(s), under exclusive license to Springer Nature Singapore Pte Ltd. 2023
Y. Cao et al. (eds.), *Automated and Electric Vehicle: Design, Informatics and Sustainability*, Recent Advancements in Connected Autonomous
Vehicle Technologies 3, https://doi.org/10.1007/978-981-19-5751-2_1

| ECMS | Equivalent Consumption Minimization Strategy | LQR | Linear Quadratic Regulator |
| HEV | Hybrid Electric Vehicle | | |

1.1 Introduction

As one of the main travel tools, the technology in the automotive field is also developing rapidly. In the automotive field, energy management is mainly used as a multi-energy source distribution management technology, which has been widely studied and applied [1]. The traditional internal combustion engine vehicle takes advantage of the high energy density of petroleum fuel to power the vehicle for long distances. However, the internal combustion engine has the disadvantages of exhaust pollution and low fuel economy. The main reasons for the fuel economy shortage of internal combustion engine are as follows.

(1) Internal combustion engine works in a lower efficiency area under some working conditions;
(2) Loss of vehicle energy during braking.

With the development of vehicles, people further improve the working efficiency of the engine through other power sources (such as batteries to provide electric energy, motor to provide power output). Batteries and motors are added to the traditional internal combustion engine to optimize the working area of the engine, and then improve the economy. The energy distribution between motor and internal combustion engine becomes the focus and difficulty of the research. In recent years, fuel oil, natural gas and fuel cell are used as the power source to drive the automobile. Energy management technology becomes the key technology to improve the rational distribution of power source energy. The reasonable distribution of electric power (electric energy management) is the key factor to improve the economy of electric vehicles. This chapter mainly analyzes and explores energy management of multi-power source vehicles for the purpose of improving automobile economy.

1.2 Architecture

In petrol-electric hybrid vehicles, the engine and motor are the main power components to drive the vehicle, and the rational distribution of engine and motor power has become the core technology of energy management [2]. The rationality of energy distribution is beneficial to realize the decoupling of engine speed and torque, improve the working area of the engine and improve the working efficiency of the engine. At present, petrol-electric hybrid electric vehicles are mainly divided into series, parallel and hybrid. The series hybrid electric vehicle drives the generator to generate electricity and provides energy output power to the motor. This structure can achieve the decoupling of engine speed and torque. However, this structure is electrically coupled and has a large energy loss due to the energy conversion process, as shown in Fig. 1.1.

Parallel hybrid vehicle also has motor and engine two power components. They are mechanically coupled and reduce the loss of energy conversion. However, the simultaneous decoupling characteristics of engine speed and torque are limited. According to the different structure, it can also achieve the decoupling characteristics of speed as shown in Fig. 1.2. The purpose of optimizing engine performance and improving automobile economy is realized.

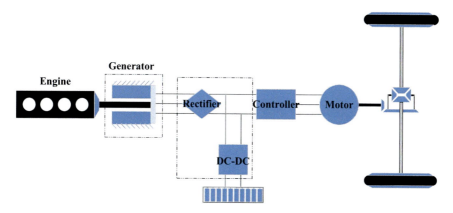

Fig. 1.1 The architecture of the series hybrid electric vehicle

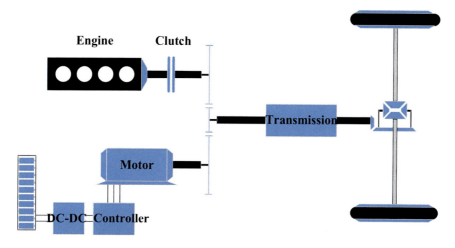

Fig. 1.2 The architecture of the parallel hybrid electric vehicle

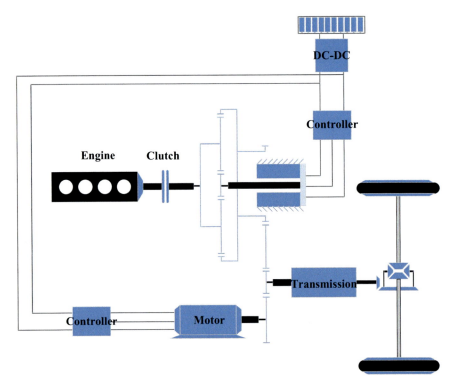

Fig. 1.3 The architecture of the hybird hybrid electric vehicle

Hybrid hybrid electric vehicle has more than two power components, combining the advantages of series and parallel. The architecture of it is shown in Fig. 1.3. The energy distribution of the power shunt type is realized by using the planetary array structure, and the working efficiency of the engine is improved. Series and parallel uses clutch to realize the switch between series and parallel, which improves the economy. However, the structure of hybrid hybrid electric vehicle is complex and the control logic is complex, so the energy distribution and output between different power components become the key and difficult point.

1.3 Rule-Based Energy Management Strategy

At present, energy management is mainly divided into two research methods: energy management control strategy based on rules and energy management control strategy based on optimization [3]. Rule-based energy management strategy mainly includes deterministic rules and fuzzy rules. It makes corresponding control strategy through a large number of experiments, experts' experience, mathematical model and other known conditions. It is simple and easy to implement, good reliability and other advantages, and is widely used [4, 5].

1.3.1 Energy Management Control Based on Deterministic Rules

1.3.1.1 Thermostat Control

Thermostat control strategy refers to the engine at a constant power output. Thermostat control strategy is most used in series petrol-electric hybrid vehicle energy management [6]. The engine of series vehicles can be decoupled (electrically coupled) from the speed and torque of the vehicle's output shaft. The engine works at the optimal working point and drive the generator to generate electricity with constant power output. Battery state of charge (SOC) is the only threshold for engine startup, as shown in Fig. 1.4.

When the battery SOC drops to a set threshold, the engine starts and outputs constant power near the optimal fuel consumption point (single point control), as shown in Table 1.1. If the output power of the engine driven generator is higher than the power required by the motor to drive the car, part of the power is used for the motor to drive the car, and the other part of the power controller controls the generator to charge the battery pack.

where is the request power; is the engine power; . P_{Best}. is the most economical operating power at current engine speed; P_b is the battery power.

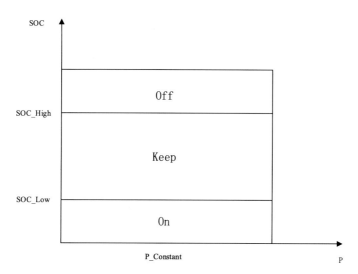

Fig. 1.4 Thermostat control

Table 1.1 Thermostat control. $SOC < SOC_{Low}$

$.~ P_r > 0~.$	$..~ P_r \leq 0~..$
$P_e = P_{Best}$ $P_b = P_r - P_{Best}$	$P_e = 0$ $|P_b| = |P_r| - |P_{Brake}|$

Table 1.2 Thermostat control. $SOC > SOC_{High}$

$P_r > 0$	$P_r \leq 0$
$P_e = 0$ $P_b = P_r$	$P_e = P_b = 0$ $P_{Brake} = P_r$

Table 1.3 Thermostat control. $SOC_{Low} < SOC < SOC_{High}$

$P_r > 0$		$P_r \leq 0$
$P_e = P_{Best}$ $P_b = P_r - P_{Best}$	$P_e = 0$ $P_b = P_r$	$P_e = 0$ $|P_b| = |P_r| - |P_{Brake}|$

When the SOC rises to the set high threshold, the engine shuts off and the battery provides all the instantaneous power requirements of the vehicle, as shown in Table 1.2. The battery provides extra power to the motor when the power generated by the engine's generator is less than what the motor needs to drive the car.

When the SOC is between the two thresholds, power components remain in the previous operating state, as shown in Table 1.3.

In thermostat control, the battery balances the power output from the engine with the power required by the motor. The control system of thermostat control strategy is simple and easy to implement. However, there are more energy conversion times and the efficiency is not high. Excessive circulation of the battery will have an adverse effect on the battery itself.

1.3.1.2 State Machine Control

Finite state machine control, referred to as state machine control, is a mathematical model consisting of multiple states. The State represents a certain property of an object [7]. By triggering the set conditions, the transition between different states can be realized. The transfer between states is realized by the logic conditions of thresholds triggered by real-time signals. The action after entering each state is to execute the corresponding control strategy. Since any two states can be transferred, only part of the transition relationship between states is established.

In the automotive field, according to different optimization purposes, through a large number of experiments and experts' experience, the vehicle is divided into different operating states. Different control thresholds are used to transfer the states between each operating state. For example, most fuel cell vehicles use state machine

1 Energy Efficient Control of Vehicles

control to coordinate the energy distribution between different power sources and improve the economy of the vehicle. The control strategy is simple and easy to implement and widely used.

1.3.1.3 Power Following Strategy Control

The power following control strategy determines the working state of the engine according to the SOC of the battery and the load of the vehicle, which is mostly used in the series structure of petrol-electric hybrid electric vehicle. It is an optimization of the thermostat control strategy. Different working points of the engine can be determined according to the different running states of the vehicle. The power of the engine follows the power required by the car to drive. Similar to the traditional vehicle, the engine speed and torque in the series hybrid electric vehicle are decoupled from the vehicle output shaft. Compared with traditional cars, the engine speed is not directly determined by the speed, and the engine can work under the economic speed. Compared with the thermostat control strategy, the engine with power following control strategy generally operates near the optimal economic operating curve. Therefore, the power following control strategy is more adaptable to external power changes, the output power is more reasonable, reduces the number of charging and discharging cycles of the battery, and is more conducive to the protection of the battery [8]. But the output power of the engine should always follow the demand power of the vehicle, the power varies widely, and the working area of the engine becomes larger. Therefore, it is difficult to ensure that the engine works in higher fuel economy areas.

In the structure of serial-type hybrid electric vehicle, the motor is used as a direct power component for power output. The energy source of the motor (power source) is obtained by the power from the battery and the power from the generator driven by the engine. Energy management mainly distributes engine power and battery power for economic energy consumption. The power following strategy controls the start/stop working mode of the engine according to the SOC of the battery and the power of the vehicle, as shown in Fig. 1.5.

When the SOC of the battery is in low power or the power required by the vehicle is large, the engine starts to work. The engine stops working when the SOC of the battery is in a high power state and the vehicle needs a large power. In the case of other SOC and vehicle power requirements, the engine keeps the working state of the previous moment unchanged. If the output power of the engine driven generator cannot meet the demand power of the vehicle, the engine can output the maximum power under the condition of ensuring relatively economic efficiency, and the insufficient power is provided by the battery. If the required power of the vehicle is less than the minimum economic power of the engine, the engine outputs the minimum economic power, and the excess power charges the battery pack. Under different SOC and power requirements, the distribution schemes of engine power and motor power are shown in Tables 1.4, 1.5 and 1.6.

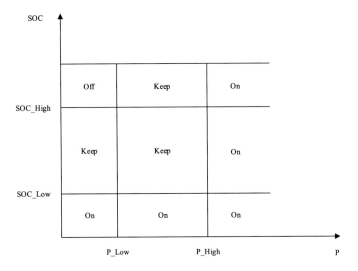

Fig. 1.5 Power following strategy control

Table 1.4 Power following strategy control. $SOC < SOC_{Low}$

P_r	P_e	P_b
$0 < P_r < P_{Low}$	$P_e = P_{Low}$	$P_b = P_e - P_r$
$P_{Low} \leq P_r \leq P_{High}$	$P_e = P_r$	$P_b = 0$
$P_{High} < P_r$	$P_e = P_r$	$P_b = 0$
$P_r \leq 0$	$P_e = 0$	$|P_b| = |P_r| - |P_{Brake}|$

Table 1.5 Power following strategy control. $SOC_{Low} \leq SOC \leq SOC_{High}$

P_r	P_e	P_b
$0 < P_r < P_{Low}$	$P_e = P_{Low}$	$P_b = P_e - P_r$
	$P_e = 0$	$P_b = P_r$
$P_{Low} \leq P_r \leq P_{High}$	$P_e = P_r$	$P_b = 0$
	$P_e = 0$	$P_b = P_r$
$P_{High} < P_r$	$P_e = P_{High}$	$P_b = P_r - P_e$
$P_r \leq 0$	$P_e = 0$	$|P_b| = |P_r| - |P_{Brake}|$

Table 1.6 Power following strategy control. $SOC_{Low} \leq SOC \leq SOC_{High}$

P_r	P_e	P_b
$0 < P_r < P_{Low}$	$P_e = 0$	$P_b = P_r$
$P_{Low} \leq P_r \leq P_{High}$	$P_e = P_r$	$P_b = 0$
	$P_e = 0$	$P_b = P_r$
$P_{High} < P_r$	$P_e = P_{High}$	$P_b = P_r - P_e$
$P_r \leq 0$	$P_e = 0$	$P_b = 0$

1.3.2 Energy Management Control Based on Fuzzy Rules

1.3.2.1 Traditional Fuzzy Logic Control

Fuzzy logic control is a kind of control method based on fuzzy set theory, fuzzy language variables and fuzzy reasoning [9]. Different from the control strategy based on the deterministic rules, fuzzy logic control strategy does not need to know the specific mathematical model of the controlled object, but relies on experience to skillfully control a complex process. When the experience is summed up in words, a qualitative and imprecise rule of control is generated. Then it is quantified into fuzzy control algorithm by fuzzy mathematics. For example, peak power SOC and vehicle power demand are described as high, medium and low. The value of the output is obtained by definite rules and fuzziness as shown in Fig. 1.6.

1.3.2.2 Adaptive Fuzzy Logic Control

Adaptive fuzzy logic control is a fuzzy logic system with adaptive learning algorithm added to the traditional fuzzy logic control technology. Its learning algorithm relies on data information to adjust the parameters of the fuzzy logic system. For example, in a certain driving cycle, the energy management effect is not very ideal, then the learning algorithm will adjust the control rules to adapt to the driving cycle. An adaptive fuzzy controller can be composed of a single adaptive fuzzy system or several adaptive fuzzy systems. Adaptive fuzzy control has two different forms:

(1) Direct adaptive fuzzy control: according to the actual performance of the system and the deviation between ideal performance, through a certain method to adjust the parameters of the controller directly;

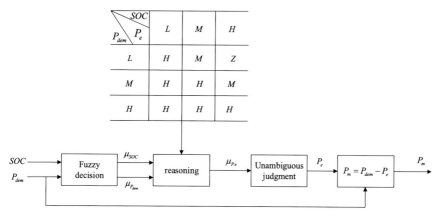

Fig. 1.6 Traditional fuzzy logic control

(2) Indirect adaptive fuzzy control: the model of the control object is obtained by on-line identification, and the fuzzy controller is designed according to the model.

1.3.2.3 Predictive Fuzzy Logic Control

Predictive fuzzy logic control strategy is organically combined with fuzzy logic control and predictive control, which can better adapt to the needs of complex process control. Its algorithm basically revolves around two directions [7].

(1) Taking process prediction information as the core, fuzzy identification and modeling methods are introduced into conventional predictive control. That is, fuzzy technology is introduced into the information processing link of the prediction model, so as to constitute the predictive fuzzy logic control. It includes using fuzzy modeling method to build object prediction model and using fuzzy technology to compensate prediction error and control law. The former uses the fuzzy model as the prediction model to improve the accuracy of predicting the output of complex objects, which is beneficial to improve the stability and robustness of the prediction algorithm. The latter uses fuzzy reasoning to compensate the output deviation of the model. And the advantage of fuzzy reasoning in uncertain information processing is used to make up the deficiency of traditional predictive control algorithm in system information processing.

(2) Taking fuzzy decision optimization as the core, the membership function and control rules of traditional fuzzy controller are optimized by using the correlation principle of predictive control and self-calibration principle, so that a certain performance measurement index tends to be optimal. Because the core of predictive control is rolling optimization, the whole algorithm can be reduced to a performance optimization problem. The traditional predictive control adopts the optimization method based on linear quadratic objective function, which minimizes the objective function in the control time domain to obtain the optimal control law. However, for complex systems, the cost of this method is very large, sometimes even impossible to achieve. To some extent, fuzzy logic control is to choose a set of controller parameters to make the controller output close to the optimal control law. Therefore, fuzzy decision can be introduced into predictive control algorithm, and various predictive fuzzy control algorithms based on fuzzy decision optimization can be obtained.

1.4 Optimization-Based Energy Management Strategy

Optimization-based energy management strategy is divided into global-optimization-based energy management strategy and instantaneous-optimization-based energy management strategy [10]. The control strategy based on optimization takes into

1 Energy Efficient Control of Vehicles

account all kinds of operating conditions and needs to consider the transient and fuel economy in the whole driving cycle. Compared with the rule-based control strategy, it has better economic effect, but the calculation is large, which increases the control difficulty.

1.4.1 Global-Optimization-Based Energy Management Strategy

1.4.1.1 Dynamic Programming

The basic idea of the dynamic programming (DP) algorithm is to transform a large multi-stage problem into multiple sub-problems of the same type. By solving the optimal solution of the sub-problem, the recursive optimization of the original problem is completed. The algorithm is based on the Bellman's principle of optimality. As a multi-stage global-optimization-based energy management strategy, regardless of its past states and decisions, the remaining decisions must constitute an optimal sub-strategy. In short, any part of the sub-strategies in the optimal policy must also be optimal [11].

The algorithm is widely used for non-aftereffect problems with deterministic conditions, where decisions are made according to stages. The term " non-aftereffect " means that when the state of the system at a certain stage is known, then the change of the system state after that stage is only related to the current stage and not to all previous stages. Therefore, the solution process of the algorithm starts from the termination stage. The algorithm is solved recursively within the boundary conditions by finding the optimal sub-problem for each stage. For each stage of the sub-problem, the optimal solution of the previous sub-problem is used in the solution process. Thereby, the complexity of the algorithm is reduced and the speed of the algorithm solution is accelerated. For DP algorithms, the following aspects need to be clarified first.

(1) Determine the stage of the optimization problem. Decompose a global optimization problem into several stages of sub-problems to be solved. The amount of stages in the solution process is denoted as k.
(2) Determine state variables and control variables for the optimization problem. The state of the system at each stage is described by the state variables. The state variable at stage k is denoted as x_k. The decision that acts on the control system to change the system state is described by the control variable. The control variables when the system state is x_k are denoted as $u_k(x_k)$. In the global optimization problem with k stages, there are $k + 1$ state variables and k control variables.
(3) Determine the constraints of the optimization problem. The state variables and control variables of the system often have various constraints, including linear and nonlinear constraints, equation constraints and inequality constraints.

Therefore, in the algorithm solution process, the state and control variables must satisfy the constraints of the optimization problem. The optimal control sequence under the global is solved within the allowed range to complete the optimal control of the system.

(4) Determine the state transfer equation of the optimization problem. The state transfer equation describes the law of the system changing from the state of the current stage to the state of the next stage. By determining the state variables and control variables of the current stage, the state variables of the next stage can be obtained to realize the transfer of the system state.

(5) Determine the cost function of the optimization problem. The cost function is used to measure the impact of the control system on the system performance in a certain state. The function depends on the current state variables and control variables. At the kth stage, when the system state variable is x_k and the control variable is u_k, the cost function is denoted as $J(x_k, u_k)$.

1.4.1.2 Pontryagin Minimum Principle

Pontriagin minimum principle (PMP), also known as maximum principle, is a method to solve optimal control problems under the condition that control vector is limited [12]. That is, under finite state or input conditions, an optimal control variable is solved to transfer the system from one state to another. The mathematical description of PMP is as shown in (1.1).

$$H(x(t), u^*(t), \lambda(t), t) \leq H(x(t), u, \lambda(t), t), \forall u \in U, t \in [t_0, t_f] \quad (1.1)$$

where U is the control domain; $u^*(t)$ is the optimal control variable; $x(t)$ is the system state variable; $\lambda(t)$ is the coordination state variable.

The mathematical description of a minimum is as shown in (1.2).

$$J = \phi(x(t_f), t_f) + \int_{t_0}^{t_f} L(x(t), u(t), t)dt \quad (1.2)$$

where $L(x(t), u(t), t)$ is the instantaneous cost value related to the control variable; $\phi(x(t_f), t_f)$ represents terminal constraint.

Because the control variable has constraints, the objective function cannot be equal to 0 by taking the derivative of the objective function with respect to the control variable. To solve the minimum problem, the constraint equation is transformed into a non-constraint equation for solution by using Lagrange multiplier method. And then the Hamiltonian function in PMP is obtained. The formula is as shown in (1.3).

$$H(x(t), u(t), \lambda(t), t) = \lambda(t) f(x(t), u(t), t) + L(x(t), u(t), t) \quad (1.3)$$

1 Energy Efficient Control of Vehicles

The PMP optimization algorithm aims to find the optimal solution through calculation, so that the Hamiltonian function can obtain a minimum value in a finite set. The formula is as shown in (1.4).

$$u^*(t) = argmin H(x(t), u(t), \lambda(t), t) \tag{1.4}$$

When obtaining the optimal control quantity u of the density function, the following conditions must be met (1.5)–(1.8).

$$\dot{\lambda}(t) = -\frac{\partial H(x(t), u(t), \lambda(t), t)}{\partial x} \tag{1.5}$$

$$\dot{x}(t) = \frac{\partial H(x(t), u(t), \lambda(t), t)}{\partial \lambda} \tag{1.6}$$

$$\mathbf{x}^*(t_0) = x_0 \tag{1.7}$$

$$\mathbf{x}^*(t_f) = x_{t\,arg\,et} \tag{1.8}$$

The necessary conditions given by PMP can be used to search for optimal control alternatives, which is called extreme control. The PMP guarantees optimal control. If it exists, it must be extreme control. If the optimal control problem has only one solution, and only one extreme control quantity, then this is the optimal control solution. Even if several extremum controls are found, it is relatively easy to apply them all at once. The optimal control is then identified as the extremum that gives the lowest total cost.

In the optimal energy management problem of gasoline-electric hybrid vehicles, the state quantity is usually SOC, and the control quantity is the output power of the engine or the output power of the motor. The state SOC must be between SOC_{max} and SOC_{min}. Thus, the set of admissible states is $\Omega_{SOC}(t) = [SOC_{min}, SOC_{max}]$. A control quantity $P_{batt}(t)$ exists in a set of permitted control $U_{P_{batt}}(t) = [P_{batt,min}(t), P_{batt,max}(t)]$. The objective function is usually fuel consumption, while engine exhaust emissions, battery aging, drivability, thermal dynamics and other factors can be included. The formula is shown in (1.9).

$$J = \varphi(\mathbf{x}(t_f)) + \int_{t_0}^{t_f} L(\mathbf{x}(t), \mathbf{u}(t), t)dt \tag{1.9}$$

The Hamiltonian function of energy management problems of gasoline-electric hybrid vehicles is shown in (1.10).

$$H\left(SOC(t), P_{batt}(t), \lambda(t), P_{req}(t)\right)$$
$$= \overset{\bullet}{m}_f\left(P_{batt}(t), P_{req}(t)\right) + (\lambda(t) + w(SOC)) \cdot \overset{\bullet}{SOC}(t) \qquad (1.10)$$

The necessary conditions for the objective function to obtain the optimal control solution are shown in (1.11)–(1.16).

$$P_{batt}^*(t) = \arg \min_{P_{batt}(t) \in U_{P_{batt}}} H\left(P_{batt}(t), SOC(t), \lambda(t), P_{req}(t)\right) \qquad (1.11)$$

$$\overset{\bullet}{SOC}^*(t) = f\left(SOC^*(t), P_{batt}^*(t)\right) \qquad (1.12)$$

$$\overset{\bullet}{\lambda}^*(t) = -\left(\lambda^*(t) + w(SOC)\right)\frac{\partial f}{\partial SOC}(SOC^*, P_{batt}^*) = h(SOC^*(t), P_{batt}^*(t), \lambda^*(t))$$
$$(1.13)$$

$$SOC^*(t_0) = SOC_0 \qquad (1.14)$$

$$SOC^*(t_f) = SOC_{t\,arg\,et} \qquad (1.15)$$

$$SOC_{\min} \leq SOC^*(t) \leq SOC_{\max} \qquad (1.16)$$

(1.12) and (1.13) represent two first-order differential equations of the variables SOC and λ. Although the two-point boundary value problem is completely defined, because one of the boundary conditions is defined at the final time, it can only be solved numerically using an iterative program.

1.4.2 Instantaneous-Optimization-Based Energy Management Strategy

1.4.2.1 Equivalent Consumption Minimum Strategy

The equivalent consumption minimization strategy (ECMS) was first proposed by Paganelli (1999) [13]. ECMS is an energy management strategy of instantaneous optimization. Its essence is to equivalent the battery power consumption to fuel consumption, that is, the fuel consumption of the vehicle into the equivalent fuel consumption of the engine and motor. ECMS and PMP algorithms are very similar, and their objective functions, too. When the ECMS was initially applied to a gasoline-electric hybrid vehicle vehicle (HEV), the difference between the initial and final charged state of the battery was very small. Negligible relative to the total amount of energy used. This means that the power storage system serves only as an energy

1 Energy Efficient Control of Vehicles

buffer. Eventually all the energy comes from fuel. And the battery can be thought of as an auxiliary reversible fuel tank. Any stored electrical energy used during the battery discharge phase must be replenished later using fuel from the engine or through regenerative braking. With the development of plug-in hybrid electric vehicle (PHEV) vehicles, ECMS technology is also gradually applied to PHEV.

The principle of ECMS is to take the power battery as a virtual engine and convert the electric energy consumed into the fuel. The equivalent fuel consumption factor determines the fuel-electric conversion efficiency of the vehicle power system and plays a controlling role in the fuel economy of the vehicle, as shown in (1.17).

$$\dot{m}_{eqv}(x(t), u(t), t) = \dot{m}_f(u(t), t) + \lambda \cdot \dot{m}_m(x(t), u(t), t) \tag{1.17}$$

where $\dot{m}_{eqv}(x(t), u(t), t)$ is the instantaneous equivalent fuel consumption; $\dot{m}_f(u(t), t)$ is the instantaneous fuel consumption of the engine; $\dot{m}_m(x(t), u(t), t)$ is the instantaneous electric consumption of the motor equivalent to the fuel consumption; λ is the equivalent factor.

In driving cycles, the fuel consumption of the vehicle should be minimized to improve the fuel economy of the vehicle. Thus, the objective function expression of ECMS optimized energy control is established as shown in (1.18).

$$J(t) = \min \int_0^t \dot{m}_{eqv}(t)dt = \min \int_0^t (\dot{m}_f + \lambda \cdot \dot{m}_m)dt \tag{1.18}$$

The Hamiltonian function is established as shown in (1.19).

$$H(x(t), u(t), \lambda(t), t) = \dot{m}_f(u(t), t) + \lambda \cdot f(x(t), u(t), t) \tag{1.19}$$

where $u(t)$ is the request torque of the motor; $x(t)$ is SOC of power battery; $f(x(t), u(t), t)$ is the instantaneous change value of SOC.

To find the optimal solution through calculation, so that the Hamiltonian function can obtain a minimum value in a finite set, as shown in (1.20).

$$u^*(t) = \arg \min H(x(t), u(t), \lambda(t), t) \tag{1.20}$$

where $u^*(t)$ is the optimal solution. Synergistic state variables and state transition variables should meet (1.21)–(1.26).

$$\dot{\lambda}(t) = -\frac{\partial H}{\partial x} = -\frac{\partial}{\partial x}\dot{m}_f(u(t), t) - \lambda(t)\frac{\partial}{\partial x}f(x(t), u(t), t) \tag{1.21}$$

$$\dot{x}(t) = \frac{\partial H}{\partial \lambda} = f(x(t), u(t), t) \tag{1.22}$$

$$SOC \in [SOC_{\min}, SOC_{\max}] \tag{1.23}$$

$$P_{batt} \in [P_{batt\_min}, P_{batt\_max}] \tag{1.24}$$

$$T_m \in [T_{m\,min}, T_{m\,max}] \tag{1.25}$$

$$T_e \in [T_{e\,min}, T_{e\,max}] \tag{1.26}$$

When the demand torque of the vehicle is known, the optimal output torque sequence of the motor is calculated according to the relationship between the engine, motor and the total demand torque. The optimal output torque of the engine is obtained as shown in (1.27).

$$T_e^*(t) = T_{req}(t) - T_m^*(t) \tag{1.27}$$

where $T_{req}(t)$ is the torque of the instantaneous demand of the vehicle; $T_e^*(t)$ is the optimal allocation torque of the engine; $T_m^*(t)$ is the optimal allocation of the motor.

1.4.2.2 Model Predictive Control

In the late 1970s, the emergence of dynamic matrix control (DMC) and model predictive heuristic control (MPHC) marked the birth of model predictive control (MPC) [14]. MPC has been a great success in the process industry since its early development and attracted many scientists to apply it to other fields. With the deepening of research and application of MPC, some scholars gradually applied MPC to vehicle control field. MPC algorithm has three main components: prediction model, rolling optimization and feedback correction. These three parts are also important differences between MPC control algorithm and other controls.

(1) Prediction model

As MPC is a predictive control algorithm based on predictive model, model plays a particularly important role in MPC. However, due to the particularity of predictive control algorithm, the requirements on the model are different from other control algorithms. Predictive control emphasizes the function of the model rather than the structure of the model. Therefore, the range of prediction models available is very wide. As long as the model has the power to predict the future input and output information of the system according to the past input and output information of the system, it can be used as a prediction model. Equation of state and transfer function can realize the above functions and can be used as prediction models naturally. Nonparametric models are common in practical industrial processes, such as step response model and impulse response model. In addition, distributed parametric

1 Energy Efficient Control of Vehicles

system and nonlinear system models can also be used as prediction models as long as they meet the requirements of prediction models. Because of the diversity of prediction model forms, the model structure forms in traditional control are greatly expanded, and it is more convenient and fast to build models based on information in practical engineering applications.

(2) Rolling optimization

In predictive control, the output control sequence is determined by the optimization of the objective function. In other words, MPC is also an optimal control algorithm. Because the determination of performance indicators is to achieve the expected control objectives, the selection of performance indicators according to the actual situation needs to be emphasized. If the accuracy of control results is highly required, the minimum variance between system output and expected output can be selected as the performance indicator. If the demand for control energy is higher, the control focus can be placed on the control energy when the output varies within a certain range. But the optimization of performance index in predictive control is obviously different from the traditional optimal control. Because the traditional optimal control is based on a fixed optimization index of the whole bureau. The optimization index in predictive control is not invariable. Its manifestation is closely related to the present moment, which is a rolling finite time domain performance index. In other words, at each sampling moment, the performance indicator only covers a period of time from that moment to a certain future. Therefore, the optimization range at each sampling point is different. This kind of rolling optimization in finite time domain obviously can only obtain global suboptimal solution. However, due to its characteristics of repeated online optimization, it is convenient to deal with the uncertainty caused by model mismatch, interference and time variation. It actually ends up keeping the control optimal.

When the system model is in the form of state space as shown in (1.28) and (1.29).

$$x_{k+1} = Ax_k + Bu_k \tag{1.28}$$

$$y_{k+1} = Cx_k \tag{1.29}$$

The objective function is shown in (1.30).

$$J = \sum_{j=0}^{N} \{\|y(k+j|k)\|_{R_{zz}} + \|u(y(k+j|k))\|_{R_{uu}}\} + F(x(k+N|k)) \tag{1.30}$$

where N is the length of time of optimization calculation in rolling optimization; $F(x(k+N|k))$ is terminal cost function; $\|u(y(k+j|k))\|_{R_{uu}}$ is the weighted norm of the input term; $\|y(k+j|k)\|_{R_{zz}}$ is the weighted norm of the output item, and the specific expression is shown in (1.31).

$$\|y(k+j|k)\|_{R_{zz}} = y(k+j|k)^T R_{zz} y(k+j|k) \tag{1.31}$$

According to the prediction model, MPC controller predicts the future output information of the system within N step length after time k. Then the control sequence with length N is calculated by objective function optimization. Then the P step-long control sequence is applied to the system. The above problem is simplified to a traditional Linear quadratic regulator (LQR) problem. Therefore, the problem can be expressed as shown in (1.32)–(1.36).

$$\min_{u} J = \sum_{j=0}^{N} \{\|y(k+j|k)\|_{R_{zz}} + \|u(y(k+j|k))\|_{R_{uu}}\} \tag{1.32}$$

$$x(k+j+1|k) = Ax(k+j|k) + Bu(k+j|k) \tag{1.33}$$

$$x(k|k) \equiv x(k) \tag{1.34}$$

$$y(k+j|k) = Cx(k+j|k) \tag{1.35}$$

$$|u(k+j|k)| \leq u_m \tag{1.36}$$

Then convert the above problems into common standard optimization problems, as shown in (1.37).

$$y(k+N|k) = CA^N x(k|k) + CA^{N-1}Bu(k|k) + CBu(k+1|k) \tag{1.37}$$

By combining, extending, and distorting the above equations, the matrix form as shown below can be derived by (1.38).

$$\min_{U(k)} J = \min_{U(k)} [H_2^T U(k) + \frac{1}{2} U(k)^T H_3 U(k)] \tag{1.38}$$

And satisfy the constraint conditions as shown in (1.39).

$$\begin{bmatrix} I_N \\ -I_N \end{bmatrix} U(k) \leq u_m \tag{1.39}$$

At k sampling, a control sequence of length H_p is obtained by optimizing computation based on the current state of the system and the prediction model. Then the control sequence is applied to the system in the control interval of length H_p. Usually $H_p = 1$, that is, only the first control in the control sequence is used. At $k+1$, repeat the above steps. With the advance of sampling time, the rolling optimization and control are realized.

1 Energy Efficient Control of Vehicles

(3) Feedback correction

In order to improve the control effect, the baseline of MPC rolling optimization at each sampling point should be consistent with the actual situation. However, due to the complexity and changeability of the actual system, it is impractical to establish an accurate mathematical model of the system. And from the practical point of view, there is no need to build an extremely accurate model. On the contrary, in practice it is easy to obtain a rough model of the system to describe its dynamic characteristics. Since when a fixed prediction model is used to describe the system, many factors existing in the actual system, such as interference, time variation, nonlinearity and model mismatch, will lead to such a fixed model deviating from the real situation of the system. So it is necessary to use additional means to correct for differences. According to the knowledge of traditional control theory, feedback can effectively overcome the influence of interference and obtain closed-loop stability. MPC rolling optimization works best when it is based on feedback. MPC obtains the control sequence of given length by optimizing the objective function at the current sampling point. Then the first term of the control sequence is applied to the system as the actual control quantity. At the next sampling time, the output information of the system is first used for feedback. Correct or compensate the prediction model. Then a new round of optimization begins. This ensures that each optimization and control is based on the latest state of the system and helps to reduce the distortion of optimization datum due to interference. The important role of output information is to give the direction of prediction model correction. Therefore, MPC optimization is not only model-based optimization, but also a closed-loop optimization to improve the model. Greatly improves the control accuracy of MPC. Enhance the actual adaptability of the control object.

For example, a fairly common prediction error can be described as (1.40).

$$e(k + 1) = y(k + 1) - \hat{y}(k + 1) \tag{1.40}$$

By adopting a weighted method to correct the predicted value at the next moment, the following results can be obtained as shown in (1.41).

$$\tilde{Y}_p = Y_p + he(k + 1) \tag{1.41}$$

where $\tilde{Y}_p = [\hat{y}(k + 1), \hat{y}(k + 2), ...]^T$ is the system output predicted at the sampling point $t = (k + 1)T$ after error correction; $h = [h_1, h_2, ...]$ is the error weighting coefficient.

After correction, the initial predicted value at the next sampling time is obtained. Since the initial predicted value of the moment is used to predict the system output value of the moment, the initial predicted value of the ext sampling moment is shown in (1.42).

$$\begin{cases} y_0(k+i) = \tilde{y}(k+i+1) + h_{i+1}e(k+1) \\ \quad y_0(k+p) = \tilde{y}(k+p) + h_p e(k+1) \end{cases} \tag{1.42}$$

The closed-loop negative feedback of the system is constructed by error correction. The closed-loop negative feedback can improve the stability of the system and improve the system performance and control precision.

1.5 Conclusion

The energy management strategy of hybrid electric vehicle is a nonlinear and complex optimization problem. Researchers began to formulate the rule-based control strategy to solve this problem. Rule-based energy management strategy has the characteristics of simple and good implementation. Therefore, it is widely used in the development of vehicle control strategy. However, its adaptability to different driving cycles varies greatly because of the single control strategy. The logic threshold of control strategy should be calibrated repeatedly according to engineering experience and expert knowledge when making control strategy according to different working conditions. It is costly and difficult to achieve optimal control effect of energy management. With the development of hybrid electric vehicle technology, the rule-based control strategy has been difficult to ensure the optimal system efficiency. Therefore, researchers use the optimal control technology and keep improving it. Although good results have been achieved, the design needs to rely on known driving cycles, and it cannot be guaranteed to be optimal in other driving cycles. With the improvement of microprocessor performance, real-time optimal control strategy based on driving cycle prediction has become a research hotspot. However, the difficulty of this strategy is to predict driving states. With the development of intelligent transportation technology, the prediction of future vehicle state will be more and more accurate, which solves the difficulty of model predictive control. Therefore, model prediction combined with intelligent transportation technology will become a research hotspot.

References

1. S.H. Mohr, Ward J, Ellem G, et al., Projection of world fossil fuels by country. Fuel (2015)
2. S. Sorrell, J. Speirs, R. Bentley et al., Global oil depletion: a review of the evidence. Energy Policy **38**(9), 5290–5295 (2010)
3. M. Kebriaei, A.H. Niasar, B. Asaei, Hybrid electric vehicles: an overview, in *2015 International Conference on Connected Vehicles and Expo (ICCVE)* (IEEE, 2016)
4. S.G. Wirasingha, A. Emadi, Classification and review of control strategies for plug-in hybrid electric vehicles. IEEE Trans. Veh. Technol. **60**(1), 111–122 (2011)
5. K.T. Chau, Y.S. Wong, Overview of power management in hybrid electric vehicles. Energy Convers. Manage. **43**(15), 1953–1968 (2002)

6. T. Lundberg, J. Persson. Hybrid powertrain control—a predictive real-time energy management system for a parallel hybrid electric vehicle (2008)
7. J. Park, J. Oh, Y. Park, et al. Optimal power distribution strategy for series-parallel hybrid electric vehicles, in *International Forum on Strategic Technology* (IEEE, 2007)
8. K. Ahn, P.Y. Papalambros, Engine optimal operation lines for power-split hybrid electric vehicles. Proc. Inst. Mech. Eng., Part D: J. Automob. Eng. **223**(9), 1149–1162 (2009). https://doi.org/10.1243/09544070J-AUTO1124
9. Y. Cheng, K. Chen, C.C. Chan, A. Bouscayrol, S. Cui, Global modeling and control strategy simulation for a hybrid electric vehicle using electrical variable transmission, in *2008 IEEE Vehicle Power and Propulsion Conference*, Harbin (2008), pp. 1–5. https://doi.org/10.1109/VPPC.2008.4677794
10. M. Schulze, R. Mustafa, B. Tilch et al., Energy management in a parallel hybrid electric vehicle for different driving conditions. Sae Int. J. Alternat. Powertrains **3**(2), 193–212 (2014)
11. H.I. Dokuyucu, M. Cakmakci, Concurrent design of energy management and vehicle traction supervisory control algorithms for parallel hybrid electric vehicles. IEEE Trans. Veh. Technol. **65**(2), 555–565 (2016)
12. W. Shabbir, S.A. Evangelou, Exclusive operation strategy for the supervisory control of series hybrid electric vehicles. IEEE Trans. Control Syst. Technol. **24**(6), 1–9 (2016)
13. M.L. Zhou, D.K. Lu, W.M. Li et al., Optimized fuzzy logic control strategy for parallel hybrid electric vehicle based on genetic algorithm. Appl. Mech. Mater. **274**, 345–349 (2013)
14. Lu, Dengke, Li, et al., Fuzzy logic control approach to the energy management of parallel hybrid electric vehicles, 592–596 (2012)

Chapter 2
Battery Management System of Electric Vehicle

Yuanjian Zhang and Zhuoran Hou

Abstract Lithium-ion rechargeable cells have the highest energy density and are the standard choice for battery packs for many consumer products, from laptops to electric vehicles. While they perform superbly, they can be rather unforgiving if operated outside a generally tight safe operating area (SOA), with outcomes ranging from compromising the battery performance to outright dangerous consequences. In order to solve this problem, Battery Management System (BMS), a technology specially used to supervise battery packs, is used for the management of battery packs. The oversight that a BMS provides usually includes: Monitoring the battery, Providing battery protection, Estimating the battery's operational state, Continually optimizing battery performance, Reporting operational status to external devices. The BMS certainly has a challenging job description, and its overall complexity and oversight outreach may span many disciplines such as electrical, digital, control, thermal, and hydraulic. This chapter classifies the topology of BMS and investigates the hardware and software architecture of BMS. The main functions of BMS are introduced.

Keywords Battery Management System · Electric vehicle · Monitoring the battery · Estimating the battery's operational state

List of Abbreviations

BMS	Battery Management System	AUTOSAR	Automotive Open System Architecture
EV	Electric Vehicle	RTE	Runtime Environment
SOA	Safe Operating Area	BSW	Basic Software

Y. Zhang (✉)
Department of Aeronautical and Automotive Engineering, Loughborough University, Leicestershire, UK
e-mail: y.y.zhang@lboro.ac.uk

Z. Hou
State Key Laboratory of Automotive Simulation and Control, Jilin University, Changchun, China

© The Author(s), under exclusive license to Springer Nature Singapore Pte Ltd. 2023
Y. Cao et al. (eds.), *Automated and Electric Vehicle: Design, Informatics and Sustainability*, Recent Advancements in Connected Autonomous Vehicle Technologies 3, https://doi.org/10.1007/978-981-19-5751-2_2

BDU	Battery Disconnect Unit	ASWC	Atomic Software Component
SOC	State Of Charge	SOH	State of Health
DOD	Depth Of Discharge	ADC	analog-to-digital converter
OCV	Open Circuit Voltage	ODEs	ordinary differential equations
EKF	Extended Kalman Filter	UKF	Unscented Kalman Filter
CDKF	Central Difference Kalman Filter	SPKF	Sigma-point Kalman Filter
EIS	Electrochemical Impedance Spectroscopy	AC	Alternating Current

2.1 Introduction

Traditional fuel vehicles are mainly composed of four modules: engine, chassis, body and electrical. New energy vehicles have little difference in chassis, body, and electrical modules compared with traditional fuel vehicles. The main difference is that power components and energy storage equipment are gradually transformed from engine and fuel tank to electric motor and power battery. Therefore, both power batteries and electric motors are hotspots in the field of electric vehicles [1, 2].

Although the design and manufacturing technology of batteries has developed over a long period, various batteries have long been popularized in consumer electronic products. But power batteries that provide power sources for transportation are a relatively new field of research because the main differences between power batteries and ordinary batteries are as follows:

(1) Different application scenarios. The batteries that provide driving power for electric vehicles are called power batteries, including traditional lead-acid batteries, nickel-metal hydride batteries, and emerging lithium-ion power lithium batteries. Power batteries are divided into power-type power batteries mainly used in hybrid vehicles and energy-type power batteries mainly used in pure electric vehicles. lithium batteries used in consumer electronic products such as mobile phones and notebook computers are generally referred to as lithium batteries to distinguish them from Power batteries for electric vehicles.

(2) Different properties. Power battery refers to the battery that provides power for vehicles. To facilitate its use, it is generally a reusable battery. An ordinary battery is a primary battery that uses lithium metal or lithium alloy as the negative electrode material and uses a non-aqueous electrolyte solution, which is different from rechargeable lithium-ion and lithium-ion polymer batteries.

(3) Different battery capacities. In the case of new batteries, the capacity of power batteries is about 1000–1500 mAh; while the capacity of ordinary batteries is above 2000 mAh, and some can reach 3400 mAh.

(4) Different discharge power. A power battery can discharge power in just a few minutes, but ordinary batteries can't do it at all, so the discharge capacity of ordinary batteries is completely incomparable to power batteries. The biggest difference between a power battery and an ordinary battery is that its discharge power is large and its specific energy is high.

Due to the limitation of the specific energy of the power battery, electric vehicles often need to be equipped with a battery pack composed of multiple single cells to meet the driving range requirements. Although the current battery manufacturing process has narrowed the differences between individual cells, there are still differences in internal resistance, capacity, and voltage between single-cell lithium batteries. It is prone to uneven heat dissipation or excessive charging and discharging. These phenomena will lead to accelerated degradation of the battery and affect the performance of the battery. Over time, these batteries in poor working conditions are likely to be damaged in advance, and the overall life of the battery pack will be greatly shortened [3].

At the same time, the power battery is the main reason for the high cost of electric vehicles. The average service life of the power battery is 5–8 years, and its performance decreases with the increase in charging times. When the battery capacity decays to below 80% of the rated capacity, the power battery is no longer for electric vehicles [4]. However, after testing, maintenance, and reorganization, the retired batteries can still be further utilized in many fields such as energy storage, distributed photovoltaic power generation, household electricity, and low-speed electric vehicles. When the battery cannot be used in cascade, it needs to be recycled and disassembled for resource processing.

To sum up, a battery management system must be equipped for the power battery pack on the electric vehicle to ensure the economy and safety of the power battery. The battery management system improves the work efficiency and service life of the entire power battery pack through effective monitoring, protection, energy balance and fault alarms for the battery pack.

2.2 The Topologies of Battery Management System

The battery management system needs to monitor the status of the battery pack and make control decisions. The structure to implement these functions can be simple, or it can become complex with ease of use and stability. According to their topology, they can be divided into the following categories.

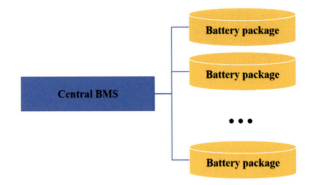

Fig. 2.1 The structure of a centralized BMS

2.2.1 Centralized BMS Architecture

The structure of the centralized BMS is shown in Fig. 2.1. All battery packs are controlled through a central BMS. The advantage of this arrangement is that the structure is compact, and there is only one BMS, so the cost is low. However, because there is only one BMS, the control precision of this arrangement is not enough. At the same time, each battery pack needs to be connected to the BMS, which also leads to complicated wiring, so the maintenance of the centralized BMS is difficult.

2.2.2 Modular BMS Topology

The structure of the modular BMS is shown in Fig. 2.2. It can be said that the modular BMS is a refined version of the centralized BMS. It is still a BMS that is directly connected to multiple battery packs as a module, and then multiple such modules are connected together. There is a primary BMS module to monitor other sub-modules. This arrangement of multiple modules, first and foremost, is the increase in the use of BMS, which inevitably leads to an increase in cost. But there are fewer battery packs connected to each BMS, which is more conducive to maintenance.

2.2.3 Primary/Subordinate BMS

The structure of the Primary/Subordinate BMS is shown in Fig. 2.3. Primary/Subordinate BMS is an upgraded structure of modular BMS. Unlike the modular BMS that connects the battery pack through multiple sub-module BMSs with the same function, the slave station of the Primary/Subordinate BMS is more limited to relaying measurement information, while the master station is dedicated to

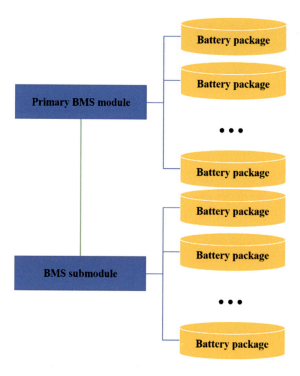

Fig. 2.2 The structure of a Modular BMS

calculation and control and external communication. Because the functional requirements of the slave station become lower, the cost of the Primary/Subordinate BMS will be lower than that of the modular BMS, and it also has the advantages of the modular BMS.

2.2.4 Distributed BMS Architecture

The structure of a distributed BMS is shown in Fig. 2.4. A distributed BMS seems to be the culmination of a modular BMS, which integrates all the electronic hardware on the control board to be placed directly on the battery or module being monitored, such that a BMS controls only one battery pack, enabling more precise control. There is no need for a large number of cable connections between the BMS and the battery pack, which is convenient for layout and inconvenient for maintenance. The cost issue brought by multiple BMSs cannot be ignored either.

To sum up, centralized BMS is the most economical layout method, but it is troubled by many lines, which is inconvenient to layout and difficult to maintain. Distributed BMS is the most expensive and the easiest to install, but it has many packages and is inconvenient to maintain. Modular BMS and Primary/Subordinate BMS are more balanced arrangements.

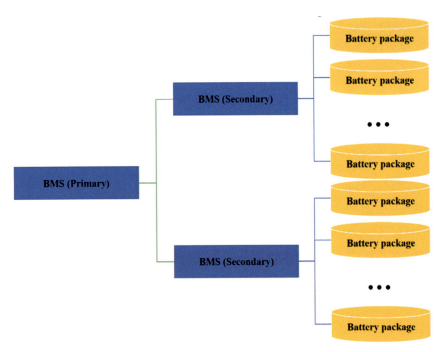

Fig. 2.3 The structure of a Primary BMS

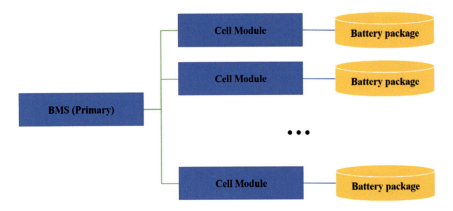

Fig. 2.4 The structure of a Distributed BMS

2.3 The Hardware of Battery Management System

If an electric vehicle is compared to the human body, then the battery system is its heart, and the BMS is the brain that controls the operation of its body. The main job of BMS is to deal with the tasks related to the onboard battery and to carry out effective monitoring, protection, energy balance and fault alarm for the battery pack. Therefore, the hardware architecture of the battery management system includes four aspects.

2.3.1 Main Board

The main board, as the brain of the BMS, collects sampling information from each slave board (usually called LCU), communicates with the vehicle through a low-voltage electrical interface, controls the action of the relay in the BDU (high-voltage breaker box), and implements battery monitoring Various states to ensure the safe use of the battery during charging and discharging.

2.3.2 The Slave Board (LCU)

The slave board (LCU), as the sentinel of the BMS, monitors the cell voltage, cell temperature and other information of the module, and transmits the information to the main board. It has the function of battery balancing. The communication method between the slave board and the main board is usually CAN, CANBUS, RS485, SMBUS, UART, and I2C.

2.3.3 Battery Disconnect Unit (BDU)

BDU is the gate for battery pack power to enter and exit. It is connected to the vehicle's high-voltage load and fast-charging harness through the high-voltage electrical interface, including pre-charging circuit, total positive relay, total negative relay, fast charging relay, etc., controlled by the main board.

2.3.4 High-Voltage Control Board

The high-voltage control board, the door guard for the battery pack's power in and out, can be integrated on the main board or independently, and it can monitor the voltage and current of the battery pack in real time. It also contains pre-charge detection and insulation detection functions.

2.4 The Software of Battery Management System

Mature BMS software development is usually based on AUTOSAR (Automotive Open System Architecture). The layered model architecture of AUTOSAR enables OEMs, suppliers, and scientific research institutions to jointly develop and cooperate efficiently to construct a powerful software system. AUTOSAR architecture divides the ECU software running on the Microcontroller into three layers: Application Layer, Runtime Environment (RTE), and Basic Software (BSW) (Fig. 2.5).

2.4.1 Application Layer

The software is divided into an Atomic Software component (ASWC), including hardware-independent application software component, Sensor Software component, Actuator Software component, etc. For the battery management system, most of the algorithm logic of its functions is performed at the application layer, which is also the core work of BMS software development.

2.4.2 Runtime Environment (RTE)

RTE provides basic communication services and supports communication between Software Component and Software Component to BSW (including program calls inside ECU, bus communication outside ECU, etc.), RTE makes the software architecture of the application layer completely separated from the specific single ECU and BSW.

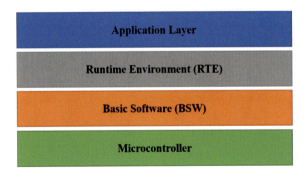

Fig. 2.5 The structure of AUTOSAR

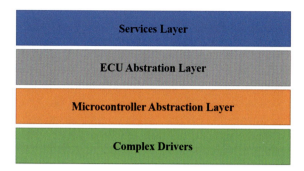

Fig. 2.6 The structure of BSW

2.4.3 Basic Software (BSW)

BSW can be subdivided into Services Layer, ECU Abstraction Layer, Microcontroller Abstraction Layer and Complex Drivers Layer (Fig. 2.6).

The Services Layer is the highest layer of the Basic Software which also applies for its relevance for the application software: while access to I/O signals is covered by the Microcontroller Abstraction Layer, the Services Layer offers Operating system services, Memory services (NVRAM management), Diagnostic Services (including KWP2000 interface and error memory), ECU state management, Logical and temporal program flow monitoring (WdgM).

The ECU Abstraction Layer is located above the Microcontroller Abstraction Layer (MCAL) and abstracts from the ECU schematic. It is implemented for a specific ECU (therefore hardware dependent) and offers an API for access to peripherals and devices regardless of their location (on-chip/off-chip) and their connection to the microcontroller (port pins, type of interface) to make higher software layers independent of the ECU hardware layout.

The Microcontroller Abstraction Layer (MCAL) is the lowest software layer of the Basic Software. It is uC dependent and contains drivers to enable the access of on-chip peripheral devices of a microcontroller and off-chip memory mapped peripheral devices by a defined API. The purpose is to make higher software layers independent of the microcontroller.

The Complex Drivers Layer can use complex drivers to allow the application layer to directly access the hardware through RTE, and can also use complex drivers to seal and convert existing non-layered software to realize the gradual implementation of the AUTOSAR software architecture.

2.5 Principal BMS Functions

The functions of the battery management system are varied, and there is no uniform standard, but it is usually related to the life and safety of the battery. To make the battery as long as possible, and as safe as possible, at least meet the regulatory requirements of various places.

Therefore, the functions that a BMS provides usually includes: (i) Monitoring the battery, (ii) Providing battery protection, (iii) Estimating the battery's operational state, (iv) Continually optimizing battery performance, (v) Reporting operational status to external devices.

2.5.1 Battery Monitoring

It mainly collects parameter information that can be directly measured by the battery during use, such as temperature, voltage and current of each battery cell, and voltage and current of the battery pack. External status detection monitors the basic status of the battery in real time and is the basis of all functions of the BMS. Without these detections, all core algorithms and application functions of the BMS are difficult to execute.

Voltage: total voltage, voltages of individual cells, or voltage of periodic taps.
Temperature: average temperature, coolant intake temperature, coolant output temperature, or temperatures of individual cells.
Coolant flow: for air or liquid cooled batteries.
Current: current in or out of the battery.

2.5.2 Battery Protection

In order to ensure the service life and safety of the battery, the battery should not be allowed to work in a dangerous working environment. The purpose of battery protection is to keep the battery within its safe operating area (SOA). The BMS controls charging and discharging and can interrupt hazardous conditions such as overvoltage, overheating, etc. to protect the battery from possible hazards.

According to the different hazardous working conditions, battery protection can be divided into two aspects: electrical protection and thermal protection. Electrical protection is further divided into current protection and voltage protection.

The electrical SOA of any battery is bounded by current and voltage. A well-designed BMS protects the battery pack by preventing the manufacturer's battery's rated voltage and current from being exceeded. At the same time, in order to prolong the battery life, the SOA range of the battery can be even further limited without affecting too much battery performance. It is also necessary to consider the situation of sudden load changes, because the use environment of electric vehicles is complex, and high peak currents for a short time may occur. In this case, the BMS must be able to tolerate the high peak demand of the battery, and also actively interrupt the current of the battery pack when the peak current lasts too long.

Chemically, lithium-ion batteries can have a wide operating temperature range. However, as a power battery for electric vehicles, it needs to meet the power and battery life of electric vehicles. In a low temperature environment, the chemical reaction speed of the battery is significantly slowed down, and the overall battery capacity will be greatly reduced, affecting the driving range of the electric vehicle. At the same time, when charging in a low temperature environment, the phenomenon of metallic lithium plating will appear on the anode, causing permanent damage, which will not only reduce the capacity, but also make the battery more prone to failure. In a high temperature environment, the battery may experience more serious thermal runaway, causing unimaginable safety problems. The BMS can control the temperature of the battery pack by heating and cooling.

2.5.3 Estimating the Battery's Operational State

During the operation of the power battery, some internal state parameters cannot be directly obtained, and can only be estimated by known external state parameters, such as:

- State of charge (SoC) or depth of discharge (DoD), to indicate the charge level of the battery.
- State of health (SoH), a variously-defined measurement of the remaining capacity of the battery as % of the original capacity.
- State of power (SoP), the amount of power available for a defined time interval given the current power usage, temperature and other conditions.
- State of Safety (SOS).
- Maximum charge current as a charge current limit (CCL).
- Maximum discharge current as a discharge current limit (DCL).
- Energy [kWh] delivered since last charge or charge cycle.
- Internal impedance of a cell (to determine open circuit voltage).
- Charge [Ah] delivered or stored (sometimes this feature is called Coulomb counter).
- Total energy delivered since first use.
- Total operating time since first use.
- Total number of cycles.
- Temperature Monitoring.
- Coolant flow for air or liquid cooled batteries.

In this book, we will focus on how to estimate SOC and SOH in Sect. 2.6.

2.5.4 Optimizing Battery Performance

On the premise that the performance of the battery itself is difficult to directly change, optimizing the operating state of the battery through BMS can enhance the usable performance of the power battery pack, prolong the service life of the power battery, and bring a better user experience. Normally, although the current power battery processing technology is relatively mature, and the manufacturing process is advanced enough, each battery of the battery pack is almost the same. However, the chemistry of the battery and the control of the battery pack are unlikely to allow every battery to have the same operating environment. Even with a brand new battery pack with well-matched cells, the cell-to-cell resemblance decreases further over time, not only due to self-discharge, but also to charge/discharge cycles, high temperatures, and general calendar aging. Under this premise, if the battery pack cannot be controlled under strict SOA conditions, the lithium-ion battery will be overcharged, and excessive electrical energy will be converted into heat energy, resulting in safety problems.

A battery pack in series with an array of cells determines the overall pack voltage, and mismatches between adjacent cells can cause a charging dilemma. As shown in Fig. 2.7 (a is a set of identical batteries, each of which is charged in the same way, so that the upper voltage threshold can be reached at the same time to cut off the charging current. While in (b, the batteries are not balanced, 3.2 The battery with V voltage will reach the charging limit in advance. In order to ensure the safety of the battery, it is necessary to cut off the charging current at this time, so that other batteries cannot be fully charged.

BMS can balance these battery differences and optimize the capacity of the battery. Here you need to define the state of charge (SOC) of the battery, which is the current available power of the battery compared to the total power when fully charged. The balancing scheme itself is usually divided into two categories: passive and active. Among them, passive balancing is the easiest to achieve and is also an explanation of the general concept of balancing. The passive approach allows each cell in the pack to have the same charge capacity as the weakest cell. It uses relatively low current to transfer a small amount of energy from high SOC cells during the charge cycle so that all cells are charged to their maximum SOC. Figure 2.8 illustrates how the BMS accomplishes this. When unbalanced, the battery can only be charged from the low limit to the high limit to be safe. The BMS monitors each cell, and when the BMS senses that a given cell is approaching its charge limit, it directs excess current around it in a top-down fashion to the next cell below. In summary, the BMS balances the battery pack in one of the following ways:

- Removes charge from the most charged battery, leaving room for additional charge current to prevent overcharging and allowing less charged batteries to receive more charge current
- Redirects some or almost all of the charging current around the most charged battery, allowing the less charged battery to receive charging current for a longer period of time.

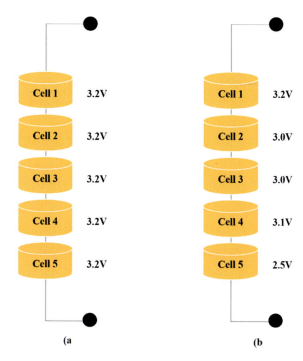

Fig. 2.7 **a** is the balanced 20 V battery stack (When each cell charged to 4.0 V cut-off), **b** is the unbalanced 20 V battery stack (May reach 18.8 V when fully chged)

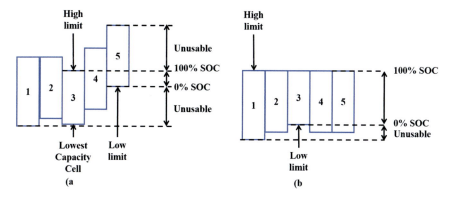

Fig. 2.8 **a** is the unbalanced battery endpoints, **b** is the balanced battery endpoints

2.5.5 *Reporting Operational Status*

The central controller of the BMS communicates internally with its hardware operating at the cell level, or externally with advanced hardware such as a laptop or HMI.

High-level external communication is simple and can be done in several ways:

- Different types of serial communication.
- CAN bus communication commonly used in the automotive environment.
- Different types of wireless communication.

Low-voltage centralized battery management systems typically do not have any internal communication. They measure the battery voltage by dividing the voltage across the resistor.

A distributed or modular battery management system must use some underlying internal unit controller (modular architecture) or controller-controller (distributed architecture) communication. These types of communications are very difficult, especially for high voltage systems. The problem is the voltage offset between cells. The xxx cell ground signal can be hundreds of volts higher than the other cell ground signal. In addition to software protocols, there are two known means of hardware communication for voltage conversion systems: opto-isolators and wireless communication. Another limitation of internal communication is the number of xxx units. For modular architectures, most hardware is limited to a maximum of 255 nodes. For high voltage systems, the seek time of all batteries is another limitation that limits the minimum bus speed and loses some hardware options. The cost of a modular system is important because it can be comparable to battery prices. A combination of hardware and software limitations are several options for internal communication:

- Isolated Serial Communication
- Wireless serial communication.

2.6 The Core Algorithm of BMS

By calculating SOC, the BMS can control charging and discharging, balancing the battery pack, thereby protecting the battery from premature capacity loss and extending its useful life. Estimating SOH is another important function of BMS, which can help users understand the usage status of the battery and is also an important parameter for estimating the remaining life of the battery.

2.6.1 Definition of SOC and SOH

The SOC and SOH are the key metrics that describe the performance of the battery and help predict its future behaviour. The SOC shows the amount of electric charge left in the battery. It's expressed as a percentage that ranges from 0 to 100% depending on the charge level. By looking at the state-of-charge indicator, a user knows the resources and understands when the battery needs to be recharged [5].

The SOC is closely tied to the battery's capacity and can also be determined as the ratio of the remaining capacity Q to the maximum capacity $Q_{\max}$.

$$SOC = Q/Q_{max} \times 100\% \tag{2.1}$$

Every cell ages and degrades, and as a result, the SOH falls below its initial level. A battery with a low SOH will discharge much faster than a battery with a high SOH. This is the result of a drop in maximum capacity—an inevitable process over time [6]. Here, SOH is defined as the ratio of current maximum available capacity Q_{cur} over the nominal value Q_{nom}, as:

$$SOH = Q_{cur}/Q_{nom} \times 100\% \tag{2.2}$$

In order to give full play to the power performance of the battery system, improve the safety of its use, prevent battery overcharge and overdischarge, prolong the service life of lithium iron phosphate batteries, optimize driving and improve the performance of electric vehicles, the BMS system must The state of charge, or SOC, is accurately estimated. SOC is an important parameter used to describe the chargeable and discharge capacity of a battery during use.

The higher the accuracy of the SOC estimation accuracy, the higher the cruising range of the electric vehicle can be for the battery of the same capacity. High-precision SOC estimation can maximize the performance of lithium battery packs. At the same time, the precise control also means that the requirements for the power battery can be reduced and the economy can be improved.

When evaluating the SOC, you can refer to the following parameters:

- Battery chemistry
- Voltage
- Current
- Capacity
- Impedance
- Charging/discharging rate
- Temperature

Accurate SOH estimation is of great significance for the optimal performance and safe operation of lithium-ion batteries.

When calculating the SOH, you can refer to the following parameters:

- Age
- Cycle life (number of charge/discharge cycles)
- Capacity
- Internal resistance
- Energy throughput
- Temperature
- Self-discharge rate
- Voltage

Calculating SOH and SOC is a difficult task. This article will introduce several commonly used state estimation methods.

2.6.2 SOC Prediction Methods

The main difficulty in calculating SOC lies in the variety of influencing factors.

(1) The high current chargeable and discharge capacity is lower than the rated capacity

The AC impedance analysis showed that the charge exchange impedance and SEI membrane impedance of the battery continued to increase with the progress of the battery pulse discharge, and the charge exchange impedance was mainly related to the size of the contact interface between the electrode active material and the electrolyte. The growth of the SEI membrane both led to The SEI film resistance increases, which also increases the charge exchange resistance.

(2) The capacity of the battery pack will change at different temperatures

Lithium power batteries are stored under high temperature conditions, and the capacity loss is the largest; when stored under low temperature conditions, the capacity loss is the smallest. The lithium power battery has a strong charge retention capability and a wide operating temperature range. Self-discharge is mainly affected by manufacturing process, materials, and storage conditions. Self-discharge is one of the main parameters to measure the performance of lithium power batteries. The lower the storage temperature of the lithium power battery, the lower the self-discharge rate, but it should also be noted that the temperature is too low or too high, which may cause damage to the lithium power battery and make it unusable.

As the temperature of lithium power batteries increases, the reaction speed increases. Relevant studies have shown that the degradation rate of lithium power batteries doubles for every 10 °C increase in the temperature of lithium power batteries. For the same cell, when the remaining capacity is 90%, the output capacity is 300kWh at 25 °C, while the output capacity at 35 °C is only 163 kWh.

Therefore, during the use of lithium power batteries, it is necessary to avoid long-term use of batteries under high temperature conditions, especially to avoid high-rate charge–discharge cycles of lithium power batteries under high temperature conditions, which is also a key target in battery thermal management.

(3) Battery capacity decay

As the battery is used, the capacity of the battery will decay for a variety of reasons, see SOH for details.

(4) Self-discharge

The self-discharge of the power battery refers to the phenomenon of automatic discharge of the lithium power battery when it is put on the open circuit. The self-discharge of the power battery will directly reduce the output power of the power battery and reduce the capacity of the power battery. The generation of self-discharge is mainly due to the thermodynamically unstable state of the electrodes in the electrolyte, and the redox reactions of the two electrodes of the power battery occur respectively.

2 Battery Management System of Electric Vehicle

Among the two electrodes, the self-discharge of the negative electrode is dominant, and the occurrence of self-discharge causes the active material to be consumed and converted into unusable thermal energy.

The magnitude of the self-discharge rate of the power battery is determined by kinetic factors, mainly depending on the nature of the electrode material, the surface state, the composition and concentration of the electrolyte, the impurity content, etc. humidity and other factors.

(5) Consistency

Generally speaking, there are many reasons for battery pack consistency problems, including the battery pack itself, which is related to the production process and quality control of the battery, referred to as internal causes; it also includes external factors during use, such as temperature, charging and discharging. Current, charge–discharge voltage, charge–discharge rate, etc., referred to as external causes. A large number of test data show that external factors are the main reason for the rapid deterioration of battery pack consistency. The consistency of the power battery pack will directly affect the power output and endurance performance of the battery pack.

Let's take a look at the most popular methods available for most battery management systems.

2.6.2.1 Open Circuit Voltage (OCV) Method

This method rests on the variation in the battery's remaining capacity or SOC to the open-circuit voltage—voltage with the disconnected current load. The stronger the dependence between voltage and SOC, the higher level of measuring accuracy you can achieve [7].

In actual operation, the battery needs to be fully charged and then discharged at a fixed rate of discharge, and stop discharging when the battery reaches the cut-off voltage. The relationship curve between OCV and SOC is obtained according to the discharge process. When the battery is in the actual working state, the current battery SOC can be obtained by looking up the OCV-SOC relationship table according to the voltage value at both ends of the battery. Although this method is effective for all kinds of batteries, it also has its own drawbacks: first, the target battery must be allowed to stand for more than 1 h before measuring OCV, so that the electrolyte inside the battery can be evenly distributed to obtain a stable terminal voltage; second, the batteries are in different At temperature or different life periods, although the open circuit voltage is the same, the actual SOC may be quite different, and the measurement results of long-term use of this method cannot be guaranteed to be completely accurate. Therefore, the open circuit voltage method, like the discharge test method, is not suitable for battery SOC estimation in operation.

2.6.2.2 Coulomb Counting (Current Integration)

As the name suggests, this method is aimed at calculating coulombs or the quantity of electric charge, which is derived from current multiplied by the time necessary for the charge to flow. To measure the remaining capacity or SOC of a battery, you can add coulombs to the initial capacity in case of charging or take them away when you discharge the battery, as:

$$SOC_{now} = SOC_{past} - \frac{1}{C_{max}} \int_0^t \eta I d\tau$$

where SOC_{past} is the initial SOC, C_{max} is the initial capacity, I is the current value, η is the efficiency of discharge.

It is not difficult to see from the above formula, but there are errors in this estimation method, which mainly come from three aspects. First, you should know the correct measurement of the initial SOC as a reference point. You can do this by fully charging or discharging the battery. Therefore, if you want to use this technique, you need to allow for periodic resetting of the battery SOC to 100% in your BMS design [8]. Second, Current sampling causes errors, including sampling accuracy and sampling interval. More accurate sensors are needed to monitor current flow. Finally, variations in battery capacity can also cause errors. Therefore, the ampere-hour integration method only records the power in and out of the battery from the outside, but ignores the changes in the internal state of the battery. At the same time, the current measurement is inaccurate, resulting in the accumulation of SOC calculation errors, which needs to be calibrated regularly.

2.6.2.3 Kalman Filtering

The method is based on the measurement and analysis of battery input and output data such as current, voltage, temperature, internal resistance and other parameters. Based on this data, you can use the Kalman filter algorithm to build an electrical model of the battery, simulate its behavior and operating conditions, and estimate the SOC.

The essence of the algorithm is that it can make an optimal estimation of the state of a complex dynamic system according to the principle of least mean square error. The nonlinear dynamic system will be linearized into the state space model of the system in the Kalman filter method. In practical application, the system updates the state variables that need to be obtained according to the estimated value at the previous moment and the observed value at the current moment. The model of "prediction-measurement-correction" eliminates the random deviation and interference of the system.

The reliability of this method depends directly on the accuracy of the electrical model, including the mathematical equations and the parameters you choose for

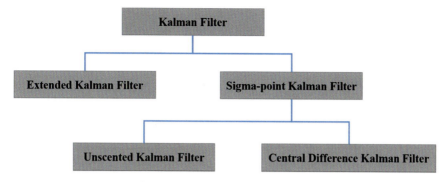

Fig. 2.9 Types of Kalman filter algorithms

those equations [9]. At the same time, the algorithm involved in this method is very complex, the calculation amount is huge, the required calculation cycle is long, and the hardware performance is demanding.

The classic Kalman filter algorithm works for systems described by linear ordinary differential equations (ODEs). For non-linear systems, you can use an extended Kalman filter (EKF) [9] that approximates non-linearity using the first-order partial derivatives. This is an upgraded version of the algorithm that needs hardware with a higher level of performance.

The unscented Kalman filter (UKF) [10] and central difference Kalman filter (CDKF) [11] are the types of a sigma-point Kalman filter (SPKF), which is another modification having higher accuracy compared to EKF. For SPKF, you solve equations for a set of points offset by sigma from the initial point (Fig. 2.9).

2.6.2.4 Alternative SOC Estimators

There are a few other techniques that can be used to estimate the SOC of a battery:

- Terminal voltage method. This method measures the reduced terminal voltage of the battery during discharge. SOC is calculated from the proportional relationship between the battery's electromotive force (EMF), its terminal voltage, and SOC. There may be serious errors in the calculation due to the sharp drop in terminal voltage at the end of the discharge.
- Impedance method. This is based on a measurement of the battery's internal impedance, which changes throughout the charge and discharge cycles. The main challenge here is the need for a way to accurately measure impedance during battery operation.
- Neural Networks. The basic process of neural network, the first step is data collection. Measurable battery parameters, such as temperature, current, and voltage data recorded by the BMS during operation, will be recorded and used as input for model training. However, not all data is related to battery aging. The second

step is to extract features that represent the aging process. The third step is to train a machine learning model to describe the relationship between the battery SOC and the extracted features. Once the model is trained, the final step is to implement it in the BMS for online application.

The application of a single SOC estimation method is rarely found in today's battery management systems. A variety of methods can be combined to achieve an accurate prediction of SOC.

2.6.3 SOH Prediction Methods

Battery SOH is caused by natural aging and cyclic aging, and is affected by internal and external factors. Over time, the battery SOH will degrade. Internal factors such as compaction pressure and drying time in the manufacturing process can cause deviations in cell aging behavior. In addition, external environmental factors, including temperature, storage SOC, charge rate, depth of discharge, mechanical stress, standby time, and run time, can significantly affect battery aging trends and aging rates. When different external environmental conditions lead to different internal states, the aging patterns of batteries can be very different, exhibiting changes in degradation trajectories such as capacity, resistance, temperature rise, etc. The aging patterns can be divided into four groups: loss of active material in the positive electrode, loss of active material in the negative electrode, loss of lithium ions, and kinetic degradation. Different types of battery aging trends can be observed even under the same external conditions. The researchers hope to provide links between external behavior and internal mechanisms under different environmental influences, revealing the influence of external factors on the battery aging path.

2.6.3.1 Internal Resistance Measurement

Internal resistance can be a telltale sign of SOH that is inversely proportional to this parameter—the higher the battery's internal resistance, the lower the SOH. Internal resistance R can be calculated through the measurements of open circuit voltage OCV and voltage V with the connected current load I. The difference between these two values will show the voltage drop. Then, you can calculate the internal resistance using Ohm's law.

$$R = (OCV - V)/I \tag{2.2}$$

However, calculating the internal resistance of the battery in this way takes time as it is OCV dependent and therefore also needs to be measured when the battery is stationary.

2.6.3.2 Internal Impedance Measurement

As a battery degrades, its internal impedance increases, just like a resistor. Therefore, impedance measurements can also be used to estimate health. At present, the main method of battery impedance measurement is electrochemical impedance spectroscopy (EIS). This method applies alternating current (AC) at different frequencies and identifies impedance as a function of frequency.

This method can calculate the internal impedance with high accuracy, thereby accurately estimating the degradation level and SOH of the battery. However, this method is computationally complex and may not be suitable for all operating conditions.

2.6.3.3 Counting Charge/Discharge Cycles

There is a relationship between the SOH of Li-ion batteries and their cycle life. Therefore, calculating the number of remaining charge/discharge cycles is the simplest SOH estimator. The cycle life claimed by the battery manufacturer can be used as a reference point, however, to calculate the number of cycles, you need to fully charge the battery and then fully discharge it, which is not consistent with the use of electric vehicles.

At the same time, some important factors such as voltage and current may affect the state of the battery, even the same battery will not have the same number of charge/discharge cycles. This method does not take into account these factors and the operating conditions of the battery.

2.6.3.4 SOC Estimators Working for SOH

The well-known techniques used for measuring the battery state-of-charge can just as well work for the SOH estimation. These include:

- **Coulomb counting**. In a battery, a reduction in state of health is equivalent to a loss of rated capacity. Therefore, once the rate of capacity decay over time is known, the SOH can be found.
- **Kalman filtering**. The Kalman filter relies on various battery parameters, including the internal resistance, which is critical for SOH estimation. At the same time, the established model can well analyze the aging mechanism of the battery.
- **Neural networks**. Neural networks are inherently data-driven and can handle both linear and nonlinear data. Therefore, the process of predicting SOC also applies to SOH, but the input and output data need to be changed.

Like with the SOC estimation, the SOH is normally defined by blending several measuring practices.

In addition, some SOH or SOC estimation methods may be able to obtain very accurate estimation results in the laboratory environment, but there are not many applications in real vehicles. For example, cycle calculation and coulomb counting methods, they all need to be fully charged or fully discharged, which is not in line with the actual application situation, and will also lead to a reduction in the life of the battery, so it is not suitable for the evaluation of electric vehicle SOH.

2.7 Conclusion

While power batteries bring fast and environmentally friendly energy, they also bring huge hidden dangers. For large lithium-ion battery packs, BMS is an indispensable component in order to ensure the safety of the battery. At the same time, BMS can also optimize the performance of the battery and prolong the service life of the battery.

References

1. L. Cai, J. Meng, D.I. Stroe, et al., An evolutionary framework for lithium-ion battery state of health estimation. J. Power Sourc. (2019)
2. Y. Li, H. Sheng, Y. Cheng, et al., Lithium-ion battery state of health monitoring based on ensemble learning, in *2019 IEEE International Instrumentation and Measurement Technology Conference (I2MTC)* (IEEE, 2019)
3. P. Guo, Z. Cheng, Y. Lei, A data-driven remaining capacity estimation approach for lithium-ion batteries based on charging health feature extraction. J. Power Sourc. (2019)
4. Y. Deng, H. Ying, E. Jiaqiang, et al., Feature parameter extraction and intelligent estimation of the State-of-Health of lithium-ion batteries. Energy 176:91–102 (2019)
5. Y. Wang, C. Zhang, Z. Chen, A method for joint estimation of state-of-charge and available energy of LiFePO$_4$ batteries. Appl. Energy **135**, 81–87 (2014)
6. Y. Shi, S. Ahmad, Q. Tong, T.M. Lim, Z. Wei, D. Ji, C.M. Eze, J. Zhao (2021) The optimization of state of charge and state of health estimation for lithium-ions battery using combined deep learning and Kalman filter methods. Int. J. Energy Res. **45**(7), 11206–11230.
7. R. Xiong, Q. Yu, C. Lin, A novel method to obtain the open circuit voltage for the state of charge of lithium ion batteries in electric vehicles by using H infinity filter. Appl. Energy **207**, 346–353 (2017)
8. J. Xie, J. Ma, K. Bai, Enhanced coulomb counting method for state-of-charge estimation of lithium-ion batteries based on peukert's law and coulombic efficiency. J Power Electron **18**(3), 910–922 (2018)
9. M. Al-Gabalawy, N.S. Hosny, J.A. Dawson et al., State of charge estimation of a Li-ion battery based on extended Kalman filtering and sensor bias. Int. J. Energy Res. **45**(5), 6708–6726 (2021)
10. D. Sun, X. Yu, C. Wang, et al. State of charge estimation for lithium-ion battery based on an intelligent adaptive extended Kalman filter with improved noise estimator (2021)
11. J. Li, H. Lu, Y. Zhou, et al. State-of-charge estimation and charge equalization for electric agricultural machinery using square-root central difference Kalman filter. IEEE (2011)

Chapter 3
Speed Forecasting Methodology and Introduction

Yuanjian Zhang and **Zhuoran Hou**

Abstract Driving prediction technique (DPT) is used to predict the distribution of various future driving conditions (FDC) such as speed, acceleration, driving behaviour etc. The quality of the prediction results has a significant impact on the performance of the corresponding predictive energy management strategy (PEMS). The quality of the prediction results has a significant impact on the performance of the corresponding predictive energy management strategy (PEMS), e.g. fuel economy (FE), battery life, etc. As vehicle speed is the most direct representation of vehicle driving characteristics, speed prediction has become a hot topic of research in China and abroad. This chapter illustrates the method to predict the vehicle speed in future domain to help intelligent vehicles grab the prior information. Besides, vehicles can use the predictor to manage the path planning and the energy management in advance to achieve the optimized control effect.

Keywords Driving prediction technique · Speed prediction · Intelligent vehicles · Intelligent transportation

List of Abbreviations

DPT	Driving Prediction Technique	FDC	Future Driving Conditions
PEMS	Predictive Energy Management Strategy	FE	Fuel Economy
LF	Left Front	FR	Right Front
RL	Left Rear	RRi	Right Rear inside
RLo	Left Rear outside	RRo	Right Rear outside

Y. Zhang (✉)
Department of Aeronautical and Automotive Engineering, Loughborough University, Leicestershire, UK
e-mail: y.y.zhang@lboro.ac.uk

Z. Hou
State Key Laboratory of Automotive Simulation and Control, Jilin University, Changchun, China

© The Author(s), under exclusive license to Springer Nature Singapore Pte Ltd. 2023
Y. Cao et al. (eds.), *Automated and Electric Vehicle: Design, Informatics and Sustainability*, Recent Advancements in Connected Autonomous Vehicle Technologies 3, https://doi.org/10.1007/978-981-19-5751-2_3

RFID	Radio Frequency Identification	MC	Markov Chain
RBF	Radical Basis Function	ANN	Artificial Neural Network
RNN	Recurrent Neural Network	LSTM	Long Short Term Memory Network

3.1 Introduction

Vehicle speed prediction refers to the use of available information to estimate the future speed of a vehicle in advance, helping the vehicle and driver to react in advance. Nowadays, speed prediction has become one of the research focuses in the field of intelligent vehicles and intelligent transportation [1], and is widely used in vehicle path planning, collision warning, advanced driver assistance systems and control strategy adjustment. However, speed prediction is highly time-varying and non-linear, which is influenced by various factors such as people, vehicles, roads and traffic, which makes speed prediction much more complex. Therefore, improving the accuracy of vehicle speed prediction has important theoretical value and wide application prospects. The current research methods for speed prediction at home and abroad can be generally divided into model-based methods and data-based methods.

3.2 Model-Driven Approach to Vehicle Speed Prediction

Model-based methods describe velocity trends using relatively accurate mathematical expressions. In general, model-based methods perform parameter calibration where the target model is known or can be established, by means of micro-modelling methods, macro-modelling methods, etc.

3.2.1 Micro-Model Based Vehicle Speed Prediction

- The longitudinal speed and acceleration of the vehicle during travel is equal to the average of these values during a certain period of time before the curve [2].
- Determination of the normal load per tire in a heavy truck with non-independent suspension using a quasi-static method in order to estimate the longitudinal speed of the vehicle when turning on a slope [3].

The following is a detailed description of the second method.

3 Speed Forecasting Methodology and Introduction

The vehicle model used in this section is a planar dynamics model describing the longitudinal and lateral dynamics of the vehicle. Before deriving the mathematical model, the following assumptions were made:

(1) The vehicle is moving forward on a road with a certain road bank angle.
(2) Consider only the longitudinal air resistance.
(3) The front and rear tires are in the same road conditions and the coefficient of friction of the ground is given in the estimation process. Therefore, the value of the tire turning stiffness is mainly influenced by the tire characteristics (including the type of tire and the number of tires on the front and rear axles).
(4) The location of the center of mass of the spring, which plays an important role in the calculation of the tire load transfer.

Climbing angle is used to describe the inclination angle of the vehicle in the forward direction, and embankment angle describes the inclination angle of the curved part of the road, This is represented in Fig. 3.1 as $\theta_{climbing}$ and φ. Basically, these two angles will cause additional longitudinal and lateral forces contributed by gravity, They vary according to the heading angle and inclination angle θ_{road}.

As shown in Fig. 3.1, the geometric relationship of the ramp can be written as:

$$\sin \theta_{climbing} = \sin \theta_{road} \cos \psi \tag{3.1}$$

$$\sin \varphi = \sin \theta_{road} \sin \psi \tag{3.2}$$

A diagram of the forces acting on the chassis tires is shown in Fig. 3.2, and defines the forces acting on the heavy truck when turning on a slope. x and y represent the vertical and horizontal axes respectively, B is the center of gravity of the entire vehicle. l_1 and l_2 are the distances from B to the front and rear axes, respectively. BF is the front wheelbase, BRi and BRo are the distance between the center of the rear inner wheel and the center of the outer wheel, respectively. m and I_{zz}. are the mass

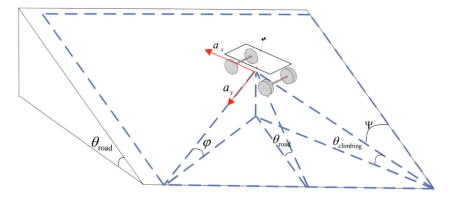

Fig. 3.1 Geometric relationship of the ramp

of the vehicle and the moment of inertia of the transverse pendulum, respectively. $FX_{FL}, FX_{FR.}, FY_{FL}, FY_{FR.}, FX_{RLo}, FX_{RLi}, FX_{RRo.}, FX_{RRi.}, FY_{RLo}, FY_{RLi.}, FY_{RRo.}, FY_{RRi}$ indicate the longitudinal and lateralorces of the tyre. The subscripts FL, FR, RL, RLo, RRi ore used to indicate left froFL), right front (FR), left rear (RL), left rear outside (RLo), right rear inside (RRi) and right rear outside (RRo) tyres. The variables u, v and r are the longitudinal speed, lateral speed and yaw rate at the centre of gravity of the truck respectively. Thus, a planar dynamics model with a certain embankment angle and non-linear tyre characteristics can be constructed as follows.

$$\frac{du}{dt} = \frac{1}{m}[(FX_{FL}\cos\delta_{FL} + FX_{FR}\cos\delta_{FR}) + (FY_{FL}\sin\delta_{FL} + FY_{FR}\sin\delta_{FR})$$
$$+FX_{RLo} + FX_{RLi} + FX_{RRo} + FX_{RRi} - F_{air} - G_x] - vr \tag{3.3}$$

$$\frac{dv}{dt} = \frac{1}{m}[(FX_{FL}\sin\delta_{FL} + FX_{FR}\sin\delta_{FR}) + (FY_{FL}\cos\delta_{FL} + FY_{FR}\cos\delta_{FR})$$
$$+FY_{RLo} + FY_{RLi} + FY_{RRo} + FY_{RRi} + G_y] - ur \tag{3.4}$$

$$\frac{dr}{dt} = \frac{1}{I_{zz}}\left[-(FY_{RLo} + FY_{RLi} + FY_{RRo} + FY_{RRi})l_2 + (FX_{RRo} - FX_{RLo})\frac{BRo}{2}\right.$$
$$+ (FX_{RRi} - FX_{RLi})\frac{BRi}{2} + (FX_{FR}\cos\delta_{FR} - FX_{FR}\sin\delta_{FR})\frac{BF}{2}$$
$$- (FX_{FL}\cos\delta_{FL} - FX_{FL}\sin\delta_{FL})\frac{BF}{2}$$
$$+(FY_{FR}\cos\delta_{FR} + FX_{FR}\sin\delta_{FR})l_1 + (FY_{FL}\cos\delta_{FL} + FX_{FL}\sin\delta_{FL})l_1]$$
$$\tag{3.5}$$

$$\frac{d\psi}{dt} = r \tag{3.6}$$

where G_x and G_y are the additional forces caused by the ramp and can be described as follows.

$$G_x = mg\sin\theta_{climbing} \tag{3.7}$$

$$G_y = mg\sin\varphi \tag{3.8}$$

The air resistance F_{air} can be obtained by the following semi-empirical formula:

$$F_{air} = sgn(u)Ac_w\frac{\rho}{2}u^2 \tag{3.9}$$

where: A is the equivalent frontal cross-sectional area of the truck, c_w is the air drag coefficient and ρ is the air density.

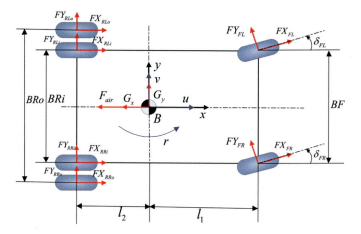

Fig. 3.2 Forces in the planar dynamics model

3.2.2 Macro-Model Based Vehicle Speed Prediction

The macroscopic traffic model for a single motorway route is a complete macroscopic model that addresses the dynamic evolution of traffic density and average traffic speed. By appropriately redefining the steady-state speed density properties normally included in the equations related to the dynamic behaviour of the average speed of traffic, taking into account the interference between vehicle classes.

The model is first formulated in a continuous space–time framework and then discretized in space and time into a system of difference equations, with the basic aggregated state variables related by the following equation:

$$q(x, t) = \rho(x, t) v(x, t) \tag{3.10}$$

where $\rho(x, t)$ is the traffic density at a given location and moment, i.e. the number of vehicles per unit length [veh/km], $v(x, t)$ is the average vehicle speed [km/h], and $q(x, t)$ is the traffic volume, i.e. the number of vehicles passing location x in a unit of time [veh/h].

In general, the macroscopic model is associated with multi-lane motorways with entrance ramps and exit ramps and describes the traffic flow as a homogeneous fluid satisfying the conservation of matter equation

$$\frac{\partial \rho(x, t)}{\partial t} + \frac{\partial q(x, t)}{\partial x} = r(x, t) - s(x, t) \tag{3.11}$$

where $r(x, t)$ and $s(x, t)$ are the inlet and outlet source terms at time t. Equation (3.11) is the basis of any macroscopic model, which is then discretised in space and time. The motorway is divided into N segments, each with at most one entrance ramp and one exit ramp, with lengths Δi, i = 1, ..., N is approximately 500–1000 m. The

Fig. 3.3 Physical structure of the highway model

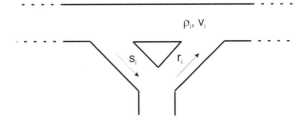

sampling interval T is chosen accordingly so as to allow a large number of vehicle lanes to be recorded. Thus, Eq. (3.11) becomes:

$$\rho_i(k+1) = \rho_i(k) + \frac{T}{\Delta_i}\left[q_{i-1}(k) - q_i(k) + r_i(k) - s_i(k)\right]$$
$$i = 1, \ldots, N \quad k = 1, 2, \ldots \tag{3.12}$$

where $\rho_i(k)$ is the traffic density of section i at time kT, $\rho_i(k)$ is the traffic volume leaving section i at time kT, $r_i(k)$ and $s_i(k)$ are the entrance ramp and the exit ramp volume of section i at time kT, respectively, as shown in Fig. 3.3.

Another aggregated variable of the model is the average traffic speed $v_i(k)$, whose dynamics are described by the following equation:

$$v_i(k+1) = v_i(k) + \frac{T}{\tau}[V(\rho_i(k)) - v_i(k)] - \frac{T}{\Delta_i}v_i(k)(v_{i-1}(k) - v_i(k))$$
$$- \frac{\nu T(\rho_{i+1}(k) - \rho_i(k))}{\tau \Delta_i(\rho_i(k) + \chi)} \quad i = 1, \ldots, N \quad k = 1, 2, \ldots \tag{3.13}$$

where τ, ν and χ are parameters determined experimentally. $V(\rho_i(k))$ is the steady-state velocity-density characteristic derived from the volume-density characteristic and is called the traffic fundamental diagram. Among the different mathematical functions for the steady-state velocity-density characteristic proposed in the literature, the most widely used expressions are:

$$V(\rho) = v_f \left\{1 - \left[\frac{\rho}{\rho_{max}}\right]^l\right\}^m \tag{3.14}$$

where the parameters v_f and ρ_{max} represent the maximum values of free speed and traffic density (congestion density) respectively, while l and m are positive real numbers.

The system consisting of Eqs. (3.12) and (3.13) is a complete classical (macro) traffic flow model. It is still necessary to define how to calculate the traffic volumes appearing in Eq. (3.12). This can be done by using the simple relation

$$q_i(k) = \rho_i(k)v_i(k) \tag{3.15}$$

3 Speed Forecasting Methodology and Introduction

A major limitation of this model, and others proposed in the literature, is that they assume that the traffic flow is a perfectly homogeneous fluid, and the unrealistic nature of this assumption is very evident. In Italy, for example, the vehicles that make up the traffic flow are divided into five categories by the motorway owners according to the purpose of the toll. As already proposed in the relevant literature for both categories of vehicles, the above model can be generalised by considering the different categories of vehicles that interact in the overall traffic flow. For this purpose, it is necessary to write model equations for each vehicle class (i.e. Equations 3.12 and 3.13). This means that each vehicle class is modelled as a traffic flow characterised by two aggregated variables related to its density and average speed. Furthermore, the interactions between the different classes of vehicles are represented by adopting suitably modified steady-state speed-density characteristics.

More precisely, it is necessary to define a new set of state variables that collect the traffic density and the average speed of traffic for the different vehicle classes. By assuming that J vehicle classes are identified, the new state variables are $\rho_{i,j}$, i = 1,..., N, j = 1,..., J and $v_{i,j}$, i = 1,..., N, j = 1,..., J, where $\rho_{i,j}$ corresponds to the density of vehicles in category j in segment i and $v_{i,j}$ is the average speed of vehicles in category j in segment i. Note that traffic volumes are now also related to each vehicle class, so the variables $q_{i,j}$, i = 1, ..., N, j = 1,..., J, the meaning of which is straightforward.

The new model equation is then

$$\rho_{i,j}(k+1) = \rho_{i,j}(k) + \frac{T}{\Delta_i}\left[q_{i-1,j}(k) - q_{i,j}(k) + r_{i,j}(k) - s_{i,j}(k)\right]$$
$$i = 1, \ldots, N \quad j = 1, \ldots, J \quad k = 1, 2, \ldots \tag{3.16}$$

$$v_{i,j}(k+1) = v_{i,j}(k) + \frac{T}{\tau}\left[V\left(\hat{\rho}_{i,j}(k)\right) - v_{i,j}(k)\right] - \frac{T}{\Delta_i}v_{i,j}(k)\left(v_{i-1,j}(k) - v_{i,j}(k)\right)$$
$$- \frac{vT\left(\rho_{i+1,j}(k) - \rho_{i,j}(k)\right)}{\tau\Delta_i\left(\rho_{i,j}(k) + \chi\right)} \quad i = 1, \ldots, N \quad j = 1, \ldots, J \quad k = 1, 2, \ldots \tag{3.17}$$

Note that the velocity-density characteristic is now the vector $\hat{\rho}_i(k) \triangleq col\left(\hat{\rho}_{i,j}(k)\right) i = 1, \ldots J, j = 1,..., N, k = 1, 2,...,$ where $\hat{\rho}_{i,j}(k)$ (k) is the traffic density of class j in segment i during the kth time interval, appropriately weighted in order to make the unit of measure uniform for all traffic densities. More precisely, the steady-state speed-density characteristic is expressed as:

$$V_{i,j}\left(\hat{\rho}_i(k)\right) = v_{f_{i,j}}\left\{1 - \left[\frac{\sum_{j=1}^{J}\hat{\rho}_{i,j}(k)}{\rho_{max_i}}\right]^{l_{i,j}}\right\}^{m_{i,j}} \tag{3.18}$$

where $v_{f_{i,j}}$, $l_{i,j}$, $m_{i,j}$ have the same meaning as in (3.14), but refer to each highway section and each vehicle class, and ρ_{max_i} is the maximum global traffic density in section i.

3.2.3 Forecasting Methods Based on Exponential Change

The exponential function-based forecasting method is to represent the forecast information in the future finite time domain in the form of exponential change, i.e. assuming that the speed in the future time domain is exponentially related to the current speed, the speed $v(k + i)$ in the future time domain at time k can be expressed as:

$$v(k + i) = v(k)(1 + \varepsilon)^i \quad i = 1, 2, \ldots N_P \qquad (3.19)$$

where $v(k)$ is the current vehicle speed (km/h), ε is the exponential coefficient and N_P is the predicted duration (s), which is usually taken as $-0.05 \sim 0.05$.

3.3 Data-Driven Approach to Vehicle Speed Prediction

Data-driven prediction methods mainly include Markov chain-based models and neural network-based models, which are based on the principle of collecting a large amount of historical working conditions data, building corresponding prediction models and combining current and historical working conditions information to achieve the prediction of vehicle speed.

3.3.1 Speed Prediction Algorithm Based on Adaptive Kalman Filtering

The core elements of the speed prediction problem of high-speed mobile vehicles based on adaptive Kalman filtering [4] in radio frequency identification (RFID) environment include: (1) the vehicle uses the RFID tag identification and the reader to read and write the status information to achieve data collection; (2) the vehicle uses the adaptive Kalman filtering algorithm to predict the speed; (3) the vehicle corrects the speed prediction value, so that the speed prediction value is close to the real value. The corresponding radio frequency identification (RFID) system model is shown in the following diagram (Fig. 3.4).

RFID systems are usually deployed on the road surface of tunnels, caves and bridges where GPS signals cannot be covered, and consist of 3 components: RFID

Fig. 3.4 Radio frequency identification (RFID) system model

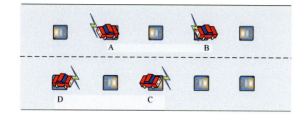

tags, readers and antennas. In the above diagram, the squares on the road surface are RFID tags that identify the vehicle attached to them and store information about the vehicle's status. The front bumper of the vehicle is equipped with readers (the two squares in front of the vehicle), which read the status information left by the vehicle in the tag and write their own status information to the tag. The status information includes the current speed of the vehicle and a time stamp. The antenna (lightning) is used to transmit the RF signal between the tag and the reader, so that the driver is aware of the relevant status information of the vehicle ahead in advance and can take better safety measures. For example, in the diagram above, vehicle A records its status information in each tag it passes, while vehicle B can obtain vehicle A's status information by reading the contents of the tag and transmitting its own status information to the next vehicle by recording it in the tag. Wherein, Vehicle A represents the front vehicle and Vehicle B represents the rear vehicle.

The entire RFID system consists of a high-speed moving vehicle and a number of RFID tags (passive tags) deployed in a large scale linear and equidistant manner. Considering the RFID system is set up on the road surface of tunnels, caves and bridge roads, for the high-speed moving vehicles, the displacement and speed are calculated as:

$$\begin{cases} x_k = x_{k-1} + v_{k-1} \cdot \Delta_t + \frac{1}{2}a_{k-1}(\Delta_t)^2 \\ v_k = v_{k-1} + a_{k-1}\Delta_t \end{cases} \quad (3.20)$$

where x_k indicates the displacement of the cart at moment k, Δ_t indicates the sampling time interval, a_{k-1} indicates the acceleration of the cart at moment k-1, and v_k indicates the velocity of the cart at moment k. In a static RFID system, with a high-speed moving cart as the research target, each tag can establish a relatively effective data transfer with each state X_k when the vehicle reaches it, and the running speed v_k of the cart and the sampling time t_k can be obtained after the data transfer. thereafter, the above equation is transformed into a state-space model. Thus, the state and measurement equations for the speed prediction problem of high-speed moving vehicles in RFID systems can be expressed as:

$$\begin{cases} X_k = AX_{k-1} + Bu_{k-1} + \omega_{k-1} \\ Z_k = CX_k + \varepsilon_k \end{cases} \quad (3.21)$$

where k = 1,2,3...N, detes the state vector of the system at moment k, A and B are state transfer matrices, u_{k-1} denotes the input to the system at moment k-1, in this case the acceleration value, ω_{k-1}. is the system error, C is the transformation matrix that converts the state to an output, Z_k denotes the state observation of the system at moment k, ε_k is the measurement error, ω_{k-1} and ε_k refer to Gaussian distributed white noise that obeys N(0, Q_{k-1}) and N(0, R_k) distributions respectively.

The core of the Kalman filter algorithm is to reconstruct the state vector of the system using the measured values, i.e. the estimated value of the previous moment and the observed value of the current moment to find the estimated value of the current moment, in order to eliminate random disturbances with the idea of "predict-measure-correct". The Kalman filter algorithm can be updated in two stages: a time update and a measurement update. The Kalman filter works as shown in the following diagram (Fig. 3.5 and Table 3.1).

When the conventional Kalman filter is applied, the parameters A, B, C, Q and R are determined by the observed system itself and by the noise during the measurement process, allowing a specific and accurate mathematical model to be obtained. However, in practice, the error covariance matrix P jingles k in the calculation of the time update due to environmental and other factors, and the measurement noise R varies depending on the vehicle being driven. Therefore, in order to better simulate the actual environment, we propose the use of adaptive Kalman filtering for speed prediction of high-speed moving vehicles.

An adaptive forgetting factor μ_k on the basis of the time update of the conventional Kalman filter algorithm is introduced, which changes the value of the error covariance matrix $\overline{P}_k$ at the time update, thus enhancing the influence of the current data and reducing the influence of the old data, keeping the data up-to-date and thus reflecting the real-time validity; at the same time, based on the state equation and the

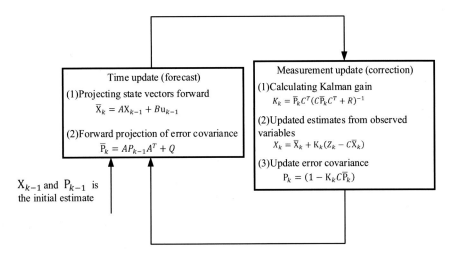

Fig. 3.5 Operating principle diagram of the Kalman filter

3 Speed Forecasting Methodology and Introduction

Table 3.1 Variables involved in the Kalman filtering algorithm

Q	Covariance matrix of the system noise
R	Covariance matrix of measurement noise
$\overline{X}_k$	A priori estimate of the state at moment k
X_k	A posteriori estimation of the state at moment k
u_{k-1}	Input value of the system at moment k-1
$\overline{P}_k$	The a priori estimation error covariance matrix at moment k
P_k	The posterior estimation error covariance matrix at moment k
Z_k	Sample values of observations at moment k
K_k	Kalman gain matrix at moment k
A,B,C	State Transfer Matrix
I	Unit matrix

measurement equation, an improved Kalman filter algorithm, namely the adaptive Kalman filter algorithm, is established.

(1) Time update (forecast)

(1) From the modified state value X_{k-1} at moment k-1 and the acceleration u_{k-1} to predict the state value $\overline{X}_k$ at moment k.

$$\overline{X}_k = AX_{k-1} + Bu_{k-1} \tag{3.22}$$

(2) From the error covariance P_{k-1} at moment k-1 to predict the error covariance P_k at moment k:

$$\overline{P}_k = \mu_k A P_{k-1} A^T + Q \tag{3.23}$$

(2) Measurement updates (forecasts)

(1) Calculate the Kalman gain K_k from the predicted error covariance $\overline{P}_k$ at moment k and the measurement noise:

$$K_k = \overline{P}_k C^T \left(C \overline{P}_k C^T + R \right)^{-1} \tag{3.24}$$

(2) Introduce the state observation Z_k of the system and use the predicted state value $\overline{X}_k$ at moment k to obtain the optimal state value X_k at the current moment:

$$X_k = \overline{X}_k + K_k \left(Z_k - C \overline{X}_k \right)^{-1} \tag{3.25}$$

(3) Update the error covariance to obtain the P_k value in preparation for predicting the new error covariance at the next moment.

$$P_{k-1} = \left(1 - K_k C \overline{P}_k\right) \tag{3.26}$$

In the RFID environment, the dynamic performance is demanding for vehicles moving at high speed. Therefore, we use a large adaptive forgetting factor μ_k to increase the strength of prediction when the parameters do not vary much, and a small adaptive forgetting factor μ_k to enhance the discrimination accuracy when the parameters vary a lot. The adaptive forgetting factor used in this section is calculated as:

$$\mu_k = max(1, tr(G_k)/tr(H_k)) \tag{3.27}$$

$$G_k = M_k - CQC^T - R \tag{3.28}$$

$$M_k = \begin{cases} 0.5 e_k e_k^T, & k = 1 \\ \frac{\mu_{k-1} e_k e_k^T}{1 + \mu_{k-1}} & k > 1 \end{cases} \tag{3.29}$$

$$e_k = Z_k - C \overline{X}_k \tag{3.30}$$

$$H_k = C A P_{k-1} A^T C^T \tag{3.31}$$

In the above equation, μ_k is the adaptive forgetting factor. When $\mu_k \leq 1$, the filter is a steady-state process; when $\mu_k > 1$, the filter may be unstable; and when $\mu_k = 1$, it can be considered as a regular Kalman filter. The adaptive forgetting factor, which is the error between the model and the actual situation and sudden changes in certain state variables, aims to limit the memory length of the Kalman filter and attenuate memory. The core idea of attenuated memory then is to apply a factor to the predicted covariance matrix, increasing the variance of the predicted state vector and thus making full use of the present measurement data. G_k and M_k denote the error variance at moment k; e_k denotes the difference between the measured and predicted values at moment k, i.e. the new interest. From the equation, it can be seen that when there is a sudden change of state, the error variance M_k becomes larger and larger as the new interest e_k increases, making the adaptive forgetting factor μ_k also larger and larger, thus improving the filter's ability to track the best state and speeding up convergence. H_k denotes the error variance at moment k, ensuring that the value of the error covariance matrix $\overline{P}_k$ satisfies symmetry and is positive definite, improving the system's Dynamic performance.

As the distance travelled increases, the error of its new interest e_k will become larger and larger, and the introduction of adaptive Kalman filtering can reduce the

3 Speed Forecasting Methodology and Introduction

impact of the new interest error. According to the improved Kalman filtering algorithm, the flow of the adaptive Kalman filtering-based speed prediction operation of high-speed moving vehicles in RFID environment is shown in Fig. 3.6.

According to the flow chart, the corresponding specific implementation steps are as follows:

1. Initialization. Initialisation of the state transfer matrices A, B, C and the acceleration u_{k-1}, the predicted and estimated covariance matrix P_{k-1}, the initial values of Q, R, and the adaptive forgetting factor μ_(k-1). where the initial state is noted as moment 0.
2. Data collection. Reads the initial vehicle data v_{k-1} and t_{k-1} collected by the tag and uses it as the initial value of the state variable.

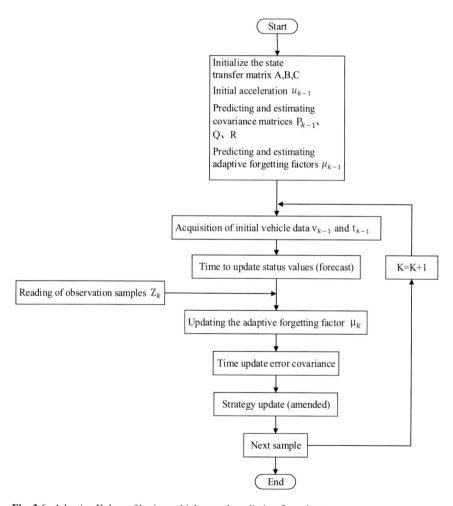

Fig. 3.6 Adaptive Kalman filtering vehicle speed prediction flow chart

3. Time update. Based on the initial state values, use Eq. (3.22) to calculate the state prediction of the system at moment 1, $\overline{X}_k$.
4. Updating the adaptive forgetting factor. Introduce the state observation Z_k of the system and use Eqs. (3.27)–(3.31) to calculate the adaptive forgetting factor μ_k of the system at moment 1.
5. Time update. Based on the initial state values and the updated adaptive forgetting factor, the error covariance matrix P_k of the system at moment 1 is calculated using equation.
6. Measurement update. Calculate the Kalman gain K_k, the optimal velocity value X_k and the error covariance P_k matrix at moment 1 according to Eqs. (3.24)–(3.26).

Proceed to the next sample. Let $K = K + 1$ and repeat steps (2)–(6) until driving out of the way of the RFID system.

3.3.2 A Markov Chain Model-Based Approach to Vehicle Speed Prediction

The future state of the vehicle is only related to the current state and not to the past state, thus it can be determined that the future state of the vehicle is a typical Markov process, so this section establishes a Markov chain model-based speed prediction method to predict the future driving conditions of the vehicle.

Markov Processes are a class of stochastic processes introduced by the Russian mathematician Andrey Markov in 1907 as an important method for studying the state space of discrete event dynamical systems. Under the condition that the state of an event is known, its future evolution does not depend on its past evolution. This property that the "future" is independent of the "past" under the condition that the "present" is known is called the Markov property, and a stochastic process with this property is called a Markov process [5]. The time parameters and state space of a Markov process can be either continuous or discrete, so a Markov Chain (MC) is defined as a class of stochastic processes that have Markovian properties and exist in a discrete set of time and state space.

A Markov chain is a set of discrete random variables with Markovian properties. Specifically, for a set $X = \{X(t), t \in T\}$ of random variables in probability space with a one-dimensional countable set as the index set, where $T = \{0,1,2,...\}$ is a discrete set of times, if the values of the random variables all lie in the countable set, i.e. $X_k = s_k$, $s_k \in s$, and the conditional probabilities of the random variables satisfy the following relation:

$$p(X_k = s_k | X_0 = s_0, X_1 = s_1, \cdots X_{k-1} = s_{k-1}) = p(X_k = s_k | X_{k-1} = s_{k-1})$$
$$(3.32)$$

3 Speed Forecasting Methodology and Introduction

Then X is called a Markov chain, the countable set $s = \{s_0, s_1, s_2 \cdots s_1\}$ is called the state space, and the values of the Markov chain in the state space are called states. The equation also shows that Markov chains are 'memoryless', i.e. the random variables at step k are conditionally independent of the rest of the random variables given the random variables at step k-1, i.e. the future state of the Markov chain X at the next moment is only related to the state at the current moment and not to the state at the previous moment.

The change in the state of a random variable in a Markov chain with respect to discrete time is called a transfer. Conditional probabilities between random variables in a Markov chain can be defined as single-step transfer probabilities as shown in Eq. (3.33) and m-step transfer probabilities as shown in Eq. (3.34).

$$P_{s_i,s_j} = p\left(X_{k+1} = s_j \,|\, X_k = s_i\right) \tag{3.33}$$

where $X_k = s_i$ and $X_{k+1} = s_j$. denote the state of the Markov chain at moment k and moment $k + 1$, respect

$$P^{(m)}_{s_i,s_j} = p\left(X_{k+m} = s_j \,|\, X_k = s_i\right) \tag{3.34}$$

where $X_k = s_i$ and $X_{k+m} = s_j$. denote the state of the Markov chain at moment k and moment $k + m$ respectively.

If the state space of a Markov chain is finite, the transfer probabilities of all states can be arranged in a matrix in a single-step transfer to obtain the single-step transfer probability matrix.

$$P_{i,j} = \left[P_{s_i,s_j} \right]_{(n+1)\times(n+1)} = \begin{bmatrix} P_{0,0} & P_{0,1} & P_{0,2} & \cdots & P_{0,n} \\ P_{1,0} & P_{1,1} & P_{1,2} & \cdots & P_{1,n} \\ P_{2,0} & P_{2,1} & P_{2,2} & \cdots & P_{2,n} \\ \vdots & \vdots & \vdots & \vdots & \vdots \\ P_{n,0} & P_{n,1} & P_{n,2} & \cdots & P_{n,n} \end{bmatrix} \tag{3.35}$$

From the properties of the probability distribution, it follows that the transfer matrix is a positive definite matrix and that the sum of the elements in each row is equal to 1:

$$\forall i, j : P_{i,j} > 0 \,\forall i : \sum_{j=0}^{n} P_{i,j} = 1 \tag{3.36}$$

The m-step transfer matrix can be defined in the same way as the single-step transfer matrix.

$$P^{(m)} = \left[P^{(m)}_{s_i,s_j} \right]_{(n+1)\times(n+1)} \tag{3.37}$$

From the properties of the m-step transfer probabilities (Chapman-Kolmogorov equation), it follows that the m-step transfer matrix can be obtained by multiplying all the transfer matrices before it, i.e.:

$$\boldsymbol{P}^{(m)} = \boldsymbol{P}^{(m-1)} \boldsymbol{P}^{(m-2)} \cdots \boldsymbol{P}^{(1)} \boldsymbol{I} \tag{3.38}$$

In the process of driving, the vehicle speed will be affected by including the driver's operation and the change of driving environment, etc. As the driver's operation intention and the change of driving environment are uncertain, thus the speed change of the vehicle in the process of driving is considered to be a discrete and Markov characteristic stochastic process, so the Markov chain model can be used to predict the vehicle speed, and its specific steps are as follows.

(1) Determining system status

Define the speed of the vehicle during travel as the system state quantity, and discretize it according to the range of values of the vehicle speed to obtain a finite set of states $V = \{V_0, V_1, V_2, \cdots, V_n\}$.

(2) Calculation of initial probabilities

For a time series $V = \{V_0, V_1, V_2, \cdots, V_n\}$, it is divided into m states, i.e. $v = \{v_0, v_1, v_2, \cdots, v_m\}$. When the last observation V_n. is unknown, the probability of v_i. occurring at the moment of V_n is calculated. Assuming that M_i of the first n known observations of the sequence V fall in state v_i., the probability of state v_i occurring is:

$$P_{v_i} = \frac{M_i}{n} \tag{3.39}$$

where P_{v_i} is referr to as the initial probability of occurrence of system state v_i.

(3) Build a single-step transfer probability matrix

A number of standard cyclic conditions are selected to form a sample database and a single-step state transfer probability matrix is calculated. Define M_{ij} as the number of transfers to state v_i at the next moment, then the single-step transfer probability of transferring state v_i to v_j is:

$$P_{v_i, v_j} = \frac{M_{ij}}{M_i} \tag{3.40}$$

In engineering applications single-step transfer probabilities can usually be obtained by statistical experiments, i.e. by approximating the probability by frequency, counting the number of adjacent state transitions that are identical and the number of occurrences of each state, i.e.

$$P_{v_i,v_j} = p\left(V_{n+1} = v_j | V_n = v_i\right) \approx \frac{Num\left(V_t = v_i \&\& V_{t+1} = v_j\right)}{Num(V_t = v_i)} \quad 0 \leq t \leq n-1$$

$$(3.41)$$

where Num denotes the number of times a condition is satisfied. After the transfer probabilities between all states have been calculated, a single-step transfer probability matrix can be built according to Eq. (3.35) for transfers between different vehicle speed states.

(4) Future vehicle speed forecast

After determining the initial probability and the single-step transfer probability matrix, the prediction of the future vehicle speed can be achieved, usually by the principle of maximum probability, that is, when the vehicle speed at the kth moment is v_i, P_{v_i,v_j} denotes the probability of the current vehicle speed v_i down to the vehicle speed v_j ($j = 0,1,2,...,$m) at the k + 1 moment, according to the maximum probability as the result of the selection as shown in the following equation:

$$V(k + 1) = v_j, \quad if \ V(k) = v_i, \ j = arg \max_{s \in I}\left(P_{v_i,v_s}\right) \quad (3.42)$$

3.3.3 Speed Prediction Based on Machine Learning

Artificial Neural Network (ANN) is a hot research topic that has emerged in the field of artificial intelligence since the 1980s. It simulates neuronal activity with a mathematical model, and is an information processing system based on mimicking the structure and function of neural networks in the brain. It is a kind of information processing system based on mimicking the structure and function of the brain's neural network. By training a large amount of sample data to automatically generalise rules, it aims to minimise the error between the actual output and the simulated output to obtain the intrinsic pattern between the sample data, so as to establish an accurate non-linear input–output mapping relationship model and achieve a wireless approximation of the strongly non-linear system [6].

When one or more adjustable parameters (weights or thresholds) of a neural network have an effect on any of the outputs, such a network is called a global approximation network. The learning speed of a global approximation network is slow because for each input, every weight on the network has to be adjusted, resulting in a slow learning speed. If only a few connected weights affect the output for a local region of the input space, then the network is called a local approximation network. Radical Basis Function (RBF) neural networks are a typical local approximation network that can handle difficult-to-resolve regularities within a system [7], have good generalisation capabilities and have a fast learning convergence rate, and have been successfully applied to non-linear function approximation, time series analysis, data classification, pattern recognition, information processing, image processing, system modeling, control and fault diagnosis, etc.

(1) RBF neural network structure

RBF neural network is a three-layer forward network with a single hidden layer, which consists of an input layer, a hidden layer and an output layer.

(1) Input layer: consists of the signal source nodes, which only serve to pass on the data information and do not transform the input information in any way. The input vector x is shown in the following equation:

$$x = [x_1, x_2, \cdots x_n]^T \tag{3.43}$$

where n is the number of neurons in the input layer.

(2) Implicit layer: consisting of multiple neurons, the kernel function (activation function) of the neurons in the implicit layer often uses a Gaussian function to transform the spatial mapping of the input information. The kernel function vector φ is.

$$\varphi = [\varphi_1, \varphi_2, \cdots \varphi_p]^T \tag{3.44}$$

where p is the number of neurons in the hidden layer.

The kernel function for each neuron takes the form:

$$\varphi_j(x) = \varphi(x - c_j) = exp\left(-\left(\frac{x - c_j}{d_j}\right)^2\right) \quad j = 1, 2 \ldots, p \tag{3.45}$$

where c_j is the central vector of the jth neuron, consisting of the central component of the jth neuron in the hidden layer corresponding to all neurons in the input layer, $c_j = [c_{j1}, c_{j2}, \cdots, c_{jn}]^T$; d_j is the width vector of the jth neuron corresponding to c_j, $d_j = [d_{j1}, d_{j2}, \cdots, d_{jn}]^T$; $x - c_j$ is the Euclidean parametrization, which can be expressed by the folling equation.

$$x - c_j = \sqrt{\sum_{i=1}^{n}(x_i - c_{ji})^2} \tag{3.46}$$

(3) Output layer: responds to the input pattern. The action function of the neurons in the output layer is a linear function, and the information output by the neurons in the hidden layer is linearly weighted and output as the output of the whole neural network. The output of the output layer neuron is:

$$y = [y_1, y_2, \cdots y_q]^T \tag{3.47}$$

3 Speed Forecasting Methodology and Introduction

$$y_k = \sum_{j=1}^{p} w_{kj}\varphi_j \quad k = 1, 2, \ldots, q \tag{3.48}$$

where q is the number of neurons in the output layer; w_{kj} is the moderation weight between the kth neuron in the output layer and the jth neuron in the hidden layer.

(2) RBF neural network training process

The main objective of RBF neural network training is to find the weights c_j. and d_j between the input and hidden layers and the weights w_j. between the hidden and output layers. The training process is divided into unsupervised and supervised learning, with the unsupervised learning method being used to determine c_j and d_j and the supervised learning method being used to determine w_j.

(1) Initialisation of weights

Initialize the central pamer $c_j = [c_{j1}, c_{j2}, \cdots, c_{jn}]^T$ for each neuron in the hidden layer, and the initial vae of the central parameter of the RBF neural network is:

$$c_{ji} = min(i) + \frac{max(i) - min(i)}{2p} + (j - 1)\frac{max(i) - min(i)}{p} \tag{3.49}$$

where min(i) and max(i) are the minimum and maximum values of all input information for the ith feature in the sample data, respectively.

Initialize the width vector of each neuron in the hidden layer $d_j = [d_{j1}, d_{j2}, \cdots, d_{jn}]^T$, the initial value of the width vector of the RBF neural network is:

$$d_{ji} = d_f \sqrt{\frac{1}{N}\sum_{k=1}^{N}(x_i^k - c_{ji})} \tag{3.50}$$

where d_f is the width adjustment factor, generally less than 1, and N is the number of input messages per neuron.

Initialize the connection weights $[w_{k1}, w_{k2}, \cdots, w_{kq}]^T$ from the hidden layer to the output layer of the RBF neural network The initial weights from the hidden layer to the output layer of the RBF neural network are:

$$w_{kj} = min(k) + j\frac{max(k) - min(k)}{q + 1} \tag{3.51}$$

where min(k). and max(k) are the minimum and maximum values of all desired outputs in the kth output neuron in the sample data, respectively.

(2) Calculate thoutput value φ_j of the jth neuron of the hidden layer

$$\varphi_j = exp\left(-\left(\frac{\|x - c_j\|}{d_j}\right)^2\right) \quad j = 1, 2 \ldots, p \quad (3.52)$$

(3) Compute the output y_k of the kth neuron of the output layer:

$$y_k = \sum_{j=1}^{p} w_{kj}\varphi_j \quad k = 1, 2, \ldots, q \quad (3.53)$$

(4) Iterative calculation of weighting parameters

The training method for the RBF neural network weight parameters was chosen as the gradient descent method. The centre, width and adjustment weight parameters were all adaptively adjusted to optimal values by learning, and the iterative process is shown in the following equation.

$$c_{ji}(t) = c_{ji}(t - 1) - \eta\frac{\partial E}{\partial c_{ji}(t - 1)} + \alpha\big[c_{ji}(t - 1) - c_{ji}(t - 2)\big] \quad (3.54)$$

$$d_{ji}(t) = d_{ji}(t - 1) - \eta\frac{\partial E}{\partial d_{ji}(t - 1)} + \alpha\big[d_{ji}(t - 1) - d_{ji}(t - 2)\big] \quad (3.55)$$

$$w_{ji}(t) = w_{ji}(t - 1) - \eta\frac{\partial E}{\partial w_{ji}(t - 1)} + \alpha\big[w_{ji}(t - 1) - w_{ji}(t - 2)\big] \quad (3.56)$$

where η is the learning factor and E is the RBF neural network objective function, which can be expressed by the following equation:

$$E = \frac{1}{2}\sum_{l=1}^{N}\sum_{k=1}^{q}(y_{lk} - o_{lk})^2 \quad (3.57)$$

where y_{lk} and o_{lk} are the network output value and the expected output value of the kth output neuron at the lth input sample.

When the RBF neural network model is used to predict future vehicle speeds, the driver driving intention and the historical speed time series are jointly used as inputs to the neural network prediction model, defined as $X_{in} = [V, H]^T$., where V denotes the historical speed series and H denotes the driving intention series, then at the kth moment we have:

$$V = V_{k-t_h}, V_{k-t_h+1}, \cdots, V_{k-1}, V_k \quad (3.58)$$

$$H = H_{k-t_h}, H_{k-t_h+1}, \cdots, H_{k-1}, H_k \quad (3.59)$$

3 Speed Forecasting Methodology and Introduction

where t_h is the length of the historical vehicle speed vector.

Define the output of the neural network prediction model at the kth moment as X_{out}, the sequence of future vehicle speeds at that moment, i.e.

$$X_{out} = V_{k+1}, V_{k+2}, V_{k+3}, \ldots, V_{k+t_p} \tag{3.60}$$

where t_p. is the leng the predicted vehicle speed vector, also known as the prediction duration.

Assuming that f_n is a non-linear mapping function for the prediction model, we have:

$$
\begin{aligned}
& V_{k+1}, V_{k+2}, V_{k+3}, \ldots, V_{k+t_p} \\
& = f_n\left(V_{k-t_h}, V_{k-t_h+1}, \ldots, V_{k-1}, V_k, H_{k-t_h}, H_{k-t_h+1}, \ldots, H_{k-1}, H_k\right) \tag{3.61}
\end{aligned}
$$

The RBF neural network is calculated and trained as described above to create a real-time speed prediction model that can be trained offline and called online.

3.3.4 Speed Prediction Based on Deep Learning

Long Short Term Memory Network (LSTM) is a special type of RNN, a neural network that can handle temporal information [8]. The future vehicle operating speed of a road is highly dependent on the vehicle operating speed in previous periods, and the prediction features selected for prediction have temporal information, so the LSTM can be used for road speed prediction.

The LSTM is an RNN network. An RNN can be thought of as multiple replications of the same neural network, with each neural network module passing messages to the next, and this structure of the RNN is suitable for processing temporal data.

The structure of an LSTM cell is shown in Fig. 3.7. The LSTM has three gates, the input gate, the forget gate and the output gate. Let h denote the output of the LSTM unit i.e. the future speed information to be predicted, c the value of the LSTM unit, X the input data i.e. the historical speed information of the vehicle, W the weight and b the bias. The candidate memory unit value $\tilde{c}_t$ at the current moment is first calculated as shown in the following equation:

$$\tilde{c}_t = tanh(W_{xc}X_t + W_{hc}h_{t-1} + b_c) \tag{3.62}$$

The output of the input gate is influenced by the previous moment's LSTM cell output value h_(t-1) in addition to the previous moment's memory cell value c_(t-1), as shown in the following equation.

$$i_t = \sigma(W_{xi}X_t + W_{hi}h_{t-1} + W_{ci}c_{t-1} + b_i) \tag{3.63}$$

Fig. 3.7 LSTM network structure diagram

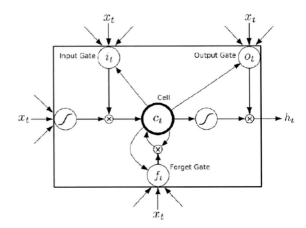

The output of the forgetting gate is influenced by the previous moment's LSTM cell output value h_{t-1} in addition to the previous moment's memory cell value c_{t-1}, as shown in the following equation.

$$f_t = \sigma\left(W_{xf}X_t + W_{hf}h_{t-1} + W_{cf}c_{t-1} + b_f\right) \quad (3.64)$$

The current state value of the memory cell c_t is shown in the following equation:

$$c_t = f_t \odot c_{t-1} + i_t \odot \tilde{c}_t \quad (3.65)$$

In the above equation, $\odot$ denotes the dot product and the memory cell state update depends on its own state and the current candidate memory cell value.

Output gate to control the output of the memory unit status value, as shown in the following equation.

$$o_t = \sigma(W_{xo}X_t + W_{ho}h_{t-1} + W_{co}c_{t-1} + b_o) \quad (3.66)$$

σ is taken as a sigmoid function.

The output value of the final LSTM cell is shown in the following equation.

$$h_t = o_t \tanh(c_t) \quad (3.67)$$

where h_t represents the future vehicle speed to be predicted, which depends on the inputs to the LSTM model, i.e. the historical vehicle speed information of the vehicle in the past, and the value of the memory unit c. The LSTM is also a form of supervised learning, so it is trained in a similar way to the RBF neural network and is not described here.

3.4 Conclusion

In summary, the model-based speed prediction method has a fast parameter learning speed and training speed, and does not require a large data set, but its fit is small and it is difficult to construct an accurate mathematical model for vehicle speeds that are highly random. Therefore, the prediction accuracy of model-based speed prediction methods still needs to be improved. The data-based method can theoretically fit highly non-linear mapping relationships and has better prediction accuracy. However, the prediction effectiveness of the method is influenced by the value of the training data, and a single piece of data can cause the model to learn unimportant mapping relationships. Therefore collecting diverse experimental data and mining the deeper value of the data will be the main research for optimising the speed prediction model.

References

1. X. Yan, C. Gu, F. Li, Z. Wang, LMP-based pricing for energy storage in local market to facilitate PV penetration. IEEE Trans. Power Syst. **12/19**, 1–1 (2017)
2. M. Magliolo, M. Bacchi, S. Sacone, et al., Traffic mean speed evaluation in freeway systems, in *Proceedings. The 7th International IEEE Conference on Intelligent Transportation Systems (IEEE Cat. No. 04TH8749)* (IEEE, 2004), pp. 233–238
3. Z. Ma, Y. Zhang, J. Yang, Velocity and normal tyre force estimation for heavy trucks based on vehicle dynamic simulation considering the road slope angle. Veh. Syst. Dyn. **54**(2), 137–167 (2016)
4. Z.H. Mir, F. Filali, An adaptive Kalman filter based traffic prediction algorithm for urban road network, in *2016 12th International Conference on Innovations in Information Technology (IIT)* (IEEE, 2016), pp. 1–6
5. J. Shin, H. Kim, S. Baek, et al., Rule-based alternator control using predicted velocity for energy management strategy. J. Dyn. Syst. Meas. Control **141**(12) (2019)
6. L.P. Zhang, W. Liu, B.N. Qi, Energy optimization of multi-mode coupling drive plug-in hybrid electric vehicles based on speed prediction. Energy **206**, 118126 (2020)
7. J. Hou, D. Yao, F. Wu et al., Online vehicle velocity prediction using an adaptive radial basis function neural network. IEEE Trans. Veh. Technol. **70**(4), 3113–3122 (2021)
8. K. Yeon, K. Min, J. Shin et al., Ego-vehicle speed prediction using a long short-term memory based recurrent neural network. Int. J. Automot. Technol. **20**(4), 713–722 (2019)

Chapter 4
Eco-Driving Behavior of Automated Vehicle

Yuanjian Zhang and **Zhuoran Hou**

Abstract Eco-driving behavior is very important in the automated vehicle, which can effectively improve the economy of the vehicle. With the development of new technologies such as the Internet of vehicles (IOVs), it is possible to obtain certain future driving conditions information and the driving state of other vehicles, and plan the economic speed of the vehicle according to the future information, thus improving the economy of the vehicle. This chapter mainly introduces the methods of speed planning. Firstly, the speed planning is described, and then the speed planning methods based on future information are introduced. In the speed planning method based on future information, the speed planning of traffic signal lights based on Internet of vehicles system and the speed planning method based on model predictive control (MPC) are mainly introduced.

Keywords Eco-driving behavior · Internet of Vehicles (IOVs) · Speed planning · Traffic signal lights · Model Predictive Control (MPC)

Y. Zhang (✉)
Department of Aeronautical and Automotive Engineering, Loughborough University, Leicestershire, UK
e-mail: y.y.zhang@lboro.ac.uk

Z. Hou
State Key Laboratory of Automotive Simulation and Control, Jilin University, Changchun, China

© The Author(s), under exclusive license to Springer Nature Singapore Pte Ltd. 2023
Y. Cao et al. (eds.), *Automated and Electric Vehicle: Design, Informatics and Sustainability*, Recent Advancements in Connected Autonomous Vehicle Technologies 3, https://doi.org/10.1007/978-981-19-5751-2_4

List of Abbreviations

IOVs	Internet of Vehicles	DSRC	Dedicated Short Range Communications
MPC	Model Predictive Control	MTFA	Microscopic Traffic Flow Analysis
V2I	Vehicle-to-Infrastructure	IDM	Intelligent Driver Model
ITS	Intelligent Transportation System	PHEV	Plug-in Hybrid EV

4.1 Introduction

With the rapid development of global economy and technology, vehicles are becoming more and more popular. At present, the global vehicle ownership is growing at the rate of tens of millions per year. Although China's auto industry started slowly, the car ownership has exceeded 150 million. Subsequently, road safety, traffic congestion, energy shortage, environmental pollution and other problems are becoming increasingly prominent [1, 2]. "Low carbon energy saving" has become the main goal of the development of the new generation of vehicles [3]. Under the background of this new round of science and technology and industry transformation, energy-saving features of new energy vehicles as the main breakthrough, focusing on the energy transform and upgrade power system, intelligent level as the main line. With advanced manufacturing and common technology such as lightweight as support, comprehensively promote car industry of low carbonization, informationization, intellectualization and high quality, is the future development direction of automobile industry in our country. The development of new energy vehicle energy-saving technology is also the main way to achieve "low carbon, information and intelligent".

Economical driving technology is an effective measure to improve vehicle fuel economy. The driving condition of the vehicle has a great influence on the economy of the vehicle. How to plan the appropriate future speed can effectively improve the economy of the vehicle, and the future driving conditions become an indispensable important information for future speed planning. In recent years, with the progress of global science and technology, the methods of obtaining future working conditions and future velocity planning have presented diversity. This chapter mainly introduces several speed planning methods, as described below.

4.2 Speed Planning of Traffic Signal Lights Based on Internet of Vehicles System

Traffic lights play an indispensable role in urban roads, traffic lights play an important part in improving traffic management and road utilization rates. However, red traffic signals can cause vehicles to changes their driving conditions. Braking and acceleration will further increase the energy consumption of the vehicle. When traffic light information is available, vehicles can take more appropriate control strategies in advance to achieve adaptive control to reduce energy consumption. The commonly used methods for traffic light information acquisition can be divided into two categories. One is the information communication method using V2I. The other is the image recognition method based on machine vision. With the development of intelligent transportation system (ITS), vehicle-to-infrastructure (V2I) technology has been used in a large number of applications to solve road traffic congestion, environmental protection and safety. In the context, vehicles are allowed to interact with neighboring infrastructure through communication protocols such as DSRC [4], LTE [5], 5G [6], etc. Drivers can take rational control through crossroads in a speed-guided manner to reduce vehicle delays and unnecessary stops. The machine vision-based image recognition method uses the vehicle's own sensors to collect traffic light images by means of color recognition or image classification. Li and Dong [7] preprocessed the collected nighttime traffic light images to achieve recognition based on regional pixel information and template matching. The experimental results showed that the recognition accuracy of the algorithm is higher than other methods. Wang et al. [8] proposed a real-time traffic light recognition system based on HDR imaging and deep learning. Experimental results show that the performance of the proposed traffic light recognition outperforms the current deep learning target detector using only bright images. Kim et al. [9] proposed a two-stage traffic light recognition method based on deep learning. Simulation results show that the proposed traffic light recognition method outperforms the traditional Faster R-CNN in terms of recognition performance. Traffic lights are an important part of urban roads. They improve traffic but limit the speed of vehicles in the airspace at the time. The image recognition method based on machine vision can only capture the current state of the traffic light. For traffic lights without countdown, the time of the remaining signal is usually unknown. All these factors limit the use of the recognition method. When utilizing traffic light information for vehicle control, it is necessary to construct an energy management strategy for the vehicle. The energy management strategy is a key technology in the design and development of vehicle control systems, which directly affects the dynamic performance and economic performance of the vehicle.

In recent years, considering the impact of traffic lights on the vehicle economy has been gradually studied. The traffic lights information based on the vehicle–road cooperative system is established to provide speed constraints for vehicles passing the green light in time–space domain. Under the constraint conditions, the energy management between the vehicle power sources is carried out, so that the power components work in an efficient area. In the whole process of optimizing the vehicle

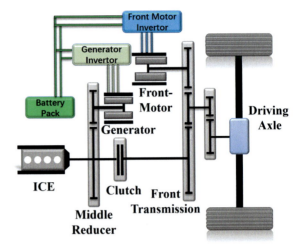

Fig. 4.1 The schematic of the PHEV configuration

through the signal intersection, the speed of the vehicle will be adjusted to avoid stopping and starting, so as to improve the economy of the vehicle. In this chapter, hybrid electric vehicle is studied to analyze the technology in detail, and its structure is shown in Fig. 4.1.

DSRC wireless communication technology is used for V2I messaging, which provides high speed data transmission, ensures low latency of the communication link and guarantees the reliability of the system. The traffic signal information is divided into timing information, phase information and localization information. The timing information includes the duration of each color of the traffic light. The timing information of all colors constitute a cycle of the traffic light. The phase information refers to the moment of the current color in a cycle. The localization information is combined with GPS to determine the distance to the next intersection. The illustration of the cooperative vehicle infrastructure system framework is shown in Fig. 4.2. The system transmits the traffic signal information to the wireless communication system through the DSRC specialized communication channel. The signal receiver on the vehicle end sends the received information to the data processing unit. The data processing unit combines the current running status of the vehicle and the traffic light information to propose constraints on the vehicle speed. Finally, a suitable energy management strategy is set. Since this paper does not consider the real-time adjustment of signal information, it does not involve the information transmitting proceeding from the vehicle to the traffic light.

In order to facilitate the analysis and sharpen the problem, the following rationalization assumptions are made in this paper:

- The time of communication transmission is ignored when the traffic information is transmitted between vehicle end and road end.
- Assume that the vehicle always drives along a straight line, ignoring steering driving conditions.

4 Eco-Driving Behavior of Automated Vehicle

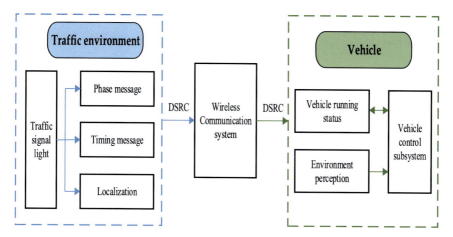

Fig. 4.2 Cooperative vehicle infrastructure system framework

- After interference from other vehicles, this is equivalent to re-specifying the initial vehicle state and solving a new optimization problem for the instantaneous-optimization-based energy management strategy. This paper studies a complete optimization process, assuming that the vehicle is not affected by other vehicles.
- Simplify the red-green-yellow signal to red-green signal for partial security assumption. The color of the traffic light is considered to be red during the period of yellow.

If the entire process of the vehicle passing is considered, there will be an infinite combination of vehicle speed and acceleration levels for a given driving situation, which will pose applicability and practicality challenges for modeling. Therefore, in order to complete the vehicle path planning and energy optimization simultaneously, this paper puts forward the following requirements for the traffic light model based on the optimized control strategy.

- The input information of the traffic light model should be consistent with the data that can be provided by the cooperative vehicle infrastructure system.
- The proposed model should be applicable to traffic lights with different phases and different distances.
- The traffic light model should give an indicator which provides constraints for subsequent energy management optimization.
- Control actions should be implemented by energy management strategies rather than the traffic light model.

Constructing the timing model of traffic lights using sine function. The formula is described as follows:

$$L(t) = \sin(\omega t + \varphi) + h \tag{4.1}$$

where ω, φ, h is the parameter of the trigonometric function, $\omega > 0$, $\varphi \in [0, 2\pi]$, $h \in (-1, 1)$ and $L(t)$ represents the lighted situation of the traffic light, respectively. If $L(t) \geq 0$ means red light at moment t. If $L(t) < 0$ means green light at moment t. The following equation is used to represent the spatial distance of the optimization task

$$D(d) = D_i - d \tag{4.2}$$

where d is the current position of the vehicle, D_i indicates the position of the next traffic light and $D(d)$ is the distance from the current vehicle position to the next traffic light position.

The more effective way to design the passing speed is to discretize the passing process and constrain the speed at each moment. By making multiple comparisons at discrete moments, the vehicle acceleration is adjusted so that the vehicle can pass through the intersection within the green light. An example diagram for replacing the dynamic adjustment process with a discrete method is shown in Fig. 4.3. The horizontal line represents a traffic light 400 m away from the starting position. The color of the horizontal line represents the color of the traffic light that changes with time. The red dots represent the position of the vehicle at different moments. The slope of the black arrow represents the speed of the vehicle at the current position. The blue line is the average speed required to ensure that the vehicle passes before the end of the green light. In Fig. 4.3, at the initial moment, the vehicle speed is slow and the black arrow is below the blue line. If the vehicle continues to drive at its current speed, it will not be able to pass the interaction before the end of the current green light. The vehicle begins to accelerate gradually until the black arrow coincides with the blue line at 30 s. The vehicle will be able to cross the intersection before the end of the green light by maintaining this speed. In the example only the velocity magnitudes of four moments are given. During the time other than these four moments, the vehicle travel process can be an arbitrary and reasonably complex acceleration and deceleration state. This allows arbitrary dynamic processes to be implemented with speed constraints at discrete moments. The method simplifies the complexity of modeling and also provides a larger optimization search space for subsequent energy management optimization.

The situation is divided into two categories combined with the process of vehicle passage. The first is when the current moment target signal is red, as shown in Fig. 4.4. The second is when the current moment target signal is green, as shown in Fig. 4.5.

In the first category of situation, in order to increase the diversity of the driving process, two subsequent green light periods are chosen to solve the vehicle constraint. As shown in the blue area in Fig. 4.4. The constrained vehicle speed is calculated by the following equation:

$$v_{\lim,\max}^j = \frac{\omega D(d)}{(2k + 2j - 1)\pi + \arcsin(h) - \varphi - \omega t} \tag{4.3}$$

4 Eco-Driving Behavior of Automated Vehicle   75

Fig. 4.3 Illustration of discretization constraints

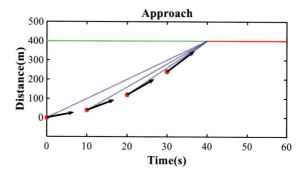

Fig. 4.4 The color of the next target traffic light is red

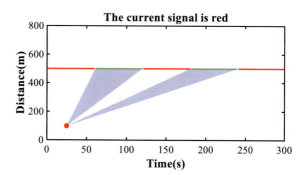

Fig. 4.5 The color of the next target traffic light is green

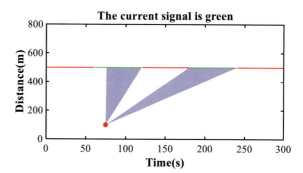

$$v_{\text{lim,min}}^{j} = \frac{\omega D(d)}{(2k+2j)\pi - \arcsin(h) - \varphi - \omega t} \quad (4.4)$$

$$k = \left[\frac{\omega t}{2\pi}\right] \quad (4.5)$$

where j represents that passing at the subsequent j th green light on, $[\cdot]$ is the downward rounding function, $v_{\text{lim,max}}^{j}$ and $v_{\text{lim,min}}^{j}$ the maximum speed constraint and the minimum speed constraint at the j th green light, respectively.

The second category of situation is shown in Fig. 4.5. It is necessary to consider whether the vehicle has time to pass during the current green light period. If it is not in time, then the speed constraint of the subsequent green light needs to be considered. The minimum value of the speed constraint and k are calculated as described in Eqs. (4.4) and (4.5). The maximum value of the speed constraint in the second category of situation is described in the following equation:

$$v_{\text{lim,max}}^1 = \infty \tag{4.6}$$

$$v_{\text{lim,max}}^2 = \frac{\omega D(d)}{(2k+3)\pi + \arcsin(h) - \varphi - \omega t} \tag{4.7}$$

Ultimately, the constraint results obtained from the traffic light model can be expressed uniformly as the following equation:

$$v_{\text{lim,min}}^j \leq v \leq v_{\text{lim,max}}^j \tag{4.8}$$

The speed limit must satisfy the relationship of the following equation:

$$v_{\text{lim,min}}^2 \leq v_{\text{lim,max}}^2 \leq v_{\text{lim,min}}^1 \leq v_{\text{lim,max}}^1 \tag{4.9}$$

Through the analysis of the above traffic signals, it can plan the reasonable speed in the future to allocate fuel and electricity reasonably, and improve the economy of the vehicle.

4.3 Speed Planning Method Based on Model Predictive Control

Model predictive control (MPC) is a feedback control algorithm that can incorporate future reference information into the control in advance to improve the performance of the controller. The MPC can also handle constraints. Any object movement can only be completed under certain conditions, not considering the constraints will lead to bad consequences. For example, the vehicle must obey the traffic laws and regulations when driving on the road, drive according to the lane and the prescribed speed, and keep a safe distance with the vehicle in front.

In order to solve the multi-objective constraint and control, nonlinear control system, the MPC control based on the simplified model, optimize and solve in the prediction of time domain and the optimal control effect in the control system which can adjust the feedback control, the control system of the rolling optimization, to enhance the robustness and stability of the control system. Because MPC performs state prediction and rolling domain optimization in finite time domain, the algorithm complexity is effectively controlled and better real-time performance is obtained.

MPC can be achieved by tracking reference speeds, which are maximum and minimum accelerations, as well as maximum and minimum speeds, estimated taking into account the influence of the environment, driving behaviour and driving safety (for example, distance from the vehicle), to provide reasonable guidance in the direction of optimization for safe and efficient driving [10]. The reference speed is generated under multiple constraints rather than through a hierarchical process, thus simplifying the implementation in real-time applications.

In this paper, we assume that velocity planning is enabled when PHEV drives along straight roads. When planning the longitudinal velocity for PHEV, the driving status is constrained not only by driving behaviors resulted from driving psychology and environment, but also by passive driving safety. To attain efficient control in PHEV with a simple framework, the estimated reference velocity should precisely reflect the driving status by incorporating safety constraints. Accordingly, an extended car-following model is preferred in this study to estimate the reference velocity. The car-following model [11] is one microscopic traffic flow analysis (MTFA) model aiming to analyze the connection between single vehicle and surrounding driving environment. The extended form of the car-following model [12] can be written as:

$$\frac{dx_j(t+\partial)}{dt} = V(\Delta x_j(t), \Delta x_{j+1}(t), \ldots, \Delta x_{j+n-1}(t)) \tag{4.10}$$

where $x_j(t+\partial)$ denotes the position of the target vehicle at time $t+\partial$, ∂ is the time lag, Δx_j is the headway of the target vehicle, j expresses the j th vehicle on the route which is chosen as the target vehicle, and n is the number of vehicles ahead of target vehicle. The top-n vehicles share the instant information including the headway detected by their radars with the target vehicle. In Eq. (4.10), the speed of the target vehicle at current position will be adjusted according to the vehicle headways to avoid collision. Therefore, the model described in Eq. (4.10) can depict vehicle driving status by incorporating safety limit. The extended car-following model can be rewritten as:

$$x_j(t+2\partial) - x_j(t+\partial) = \partial V(\Delta x_j(t), \Delta x_{j+1}(t), \ldots, \Delta x_{j+n-1}(t)) \tag{4.11}$$

By assuming that the velocity V depends on the vehicle headway, V can be formulated, as:

$$V = V\left(\sum_{h=1}^{n} B_h \Delta x_{j+h}(t)\right), h = 1, 2, \ldots, n \tag{4.12}$$

where B_h is the weight ratio on the $j+h$ th headway, and weighs the impact from forward vehicles. The impact on the target vehicle from forward vehicles will reduce with the increase of distance between them. B_h can be calculated as:

$$
B_h = \begin{cases} \dfrac{5}{6^h}, & h \neq n \\[2ex] \dfrac{1}{6^{h-1}}, & h = n \end{cases} \tag{4.13}
$$

By inserting Eq. (4.12) into Eq. (4.11), V can be rewritten as:

$$
x_j(t + 2\grave{o}) - x_j(t + \grave{o}) = \grave{o}\left(V\left(\sum_{h=1}^{n} B_h \Delta x_{j+h}(t) \right) - V\left(\sum_{h=1}^{n} B_h \Delta x_{j+h-1}(t) \right) \right) \tag{4.14}
$$

Based on Eq. (4.14) and the method in [13], V can be expressed as:

$$
\begin{aligned}
V(\Delta x_j(t), \Delta x_{j+1}(t), &\ldots, \Delta x_{j+n-1}(t)) \\
&= \frac{v_{max}}{2}\left\{ \tanh\left(\sum_{h=1}^{n} B_h \Delta x_{j+h-1}(t) - d_{hs} \right) + \tanh(d_{hs}) \right\}
\end{aligned} \tag{4.15}
$$

where d_{hs} is the safe headway under different speed, and v_{max} is the permitted maximum speed on certain route segment. The MTFA describes macroscopic behaviors of target vehicle to keep stable driving among random vehicle swarm. To this end, acceleration should be carefully constrained. The stability analysis on the macroscopic velocity calculated by Eq. (4.15) offers a valuable reference to obtain the constraints on acceleration, which, in general, are attained by considering driving safety to forward vehicles to keep driving stability within the random vehicle swarm. To keep driving safety among traffic flow, the time lag $\grave{o}$ should satisfy the following condition which is acquired by linear stability analysis [14], as:

$$
\grave{o} \gg \frac{\sum_{h=1}^{n} B_h(2h-1)}{3V'} \tag{4.16}
$$

where V' can be calculated as:

$$
V' = \frac{dV(\Delta x_i)}{d\Delta x_j} \tag{4.17}
$$

The time lag $\grave{o}$ denotes the dynamic response to reach new velocity, which is determined by driving behaviors and environment. The time lag reveals how fast driver adjusts current velocities to maintain stable driving in the random vehicle swarm. To keep safe distance to forward vehicles, driver may subjectively accelerate or decelerate in different manners, thereby reaching the same velocity with forward vehicle for stable swarm motion. Therefore, vehicle acceleration can be supposed to be proportional to the time lag. Accordingly, the minimum acceleration a_{min} can be yielded, as:

4 Eco-Driving Behavior of Automated Vehicle

$$a_{\min} \geq \frac{\gamma_t \sum_{h=1}^{n} B_h(2h-1)}{3V'} \tag{4.18}$$

where γ_t is a fixed ratio. To guarantee safety space between neighboring vehicles, the vehicle speed is constrained by:

$$v(t) \leq \frac{\Delta x_j(t) - d_{hs}}{\kappa} \tag{4.19}$$

where κ is a fixed constant. Through the derivation on Eq. (4.19), the maximum acceleration can be constrained by:

$$a_{\max} \leq \frac{d\Delta x_j(t)}{\kappa dt} \tag{4.20}$$

Note that during implementation, the maximum and minimum velocity on certain route can be acquired via IOVs [15]. The vehicle velocity in route segment should not be smaller than the permitted minimum velocity. To guarantee the implementation effect in velocity, the velocity estimation interval is set to 2 s. The exploitation of MTFA in reference velocity prediction is determined according to the application scenarios of the raised MPC based strategy. Generally, the proposed method is for longitudinal automatic driving in random vehicle fleet with stable driving status, and herein the vehicle lateral dynamic is not considered. The primary role of the forecasted velocity lies in enlightening control strategy to make optimal decisions in energy distribution within powertrain and optimize stable driving velocity within the randomly distributed fleet. There do not exist necessary constraints, such as speed limitation that is critical in specific car-following scenarios. Even so, soft constraints that ascertain vehicle can drive stably in randomly distributed vehicle swarm are still needed. Based on the analysis, MTFA, instead of intelligent driver model (IDM) [16] is applied, even the latter is more appropriate in specific car-following scenarios. The MTFA can provide rational sensing on future driving information, and disclose the interconnection between target vehicle and background vehicles without excessively addressing the strict car-following requirements.

4.4 Conclusion

As a reference of the vehicle's future driving state, the speed planning has a great influence on the vehicle's economic driving, and the quality of the speed planning directly affects the vehicle's economy. With the development of science and technology, there are more and more methods to accurately obtain the information of future driving conditions. In obtaining the information of future driving conditions, how to plan the vehicle speed has become the focus and difficulty of the research on improving the vehicle economy.

References

1. A. Alturiman, M. Alsabaan, Impact of two-way communication of traffic light signal-to-vehicle on the electric vehicle state of charge. IEEE Access **7**, 8570–8581
2. S.H. Mohr, et al., Projection of world fossil fuels by country. Fuel **141**, 120–135 (2015)
3. F.A. Wyczalek, Hybrid electric vehicles: year 2000 status. IEEE Aerosp. Electron. Syst. Mag. **16**(3), 15–25 (2001)
4. Jae, et al., Intelligent WAVE/DSRC platform technology for efficient data transmission in vehicle communication. J. Korea Multimedia Soc. **20**(9), 1519–1526 (2017)
5. S. Chen et al., LTE-V: a TD-LTE-based V2X solution for future vehicular network. IEEE Internet Things J. **3**(6), 997–1005 (2017)
6. E. Skondras, A. Michalas, D.D. Vergados, A survey on medium access control schemes for 5G vehicular cloud computing systems, in *GIIS 2018—Global Information Infrastructure and Networking Symposium* (2018).
7. J. Li, Y. Dong, A new night traffic light recognition method. J. Phys: Conf. Ser. **1176**(4), 042008 (2019)
8. J.G. Wang, L.B. Zhou, Traffic light recognition with high dynamic range imaging and deep learning. IEEE Trans. Intell. Transp. Syst. **20**(4), 1341–1352
9. H.K. Kim, et al., Traffic light recognition based on binary semantic segmentation network. Sensors (Basel, Switzerland) **19**(7) (2019)
10. H. Lim, C.C. Mi, W. Su, A distance-based two-stage ecological driving system using an estimation of distribution algorithm and model predictive control. IEEE Trans. Veh. Technol. **66**(8), 6663–6675 (2017)
11. W.X. Zhu, H.M. Zhang, Analysis of mixed traffic flow with human-driving and autonomous cars based on car-following model. Physica A—Stat. Mech. Its Appl. Article, **496**, 274–285 (2018)
12. H.X. Ge, S.Q. Dai, L.Y. Dong, Y. Xue, Stabilization effect of traffic flow in an extended car-following model based on an intelligent transportation system application. Phys. Rev. E, Article no. 066134, **70**(6), 6 (2004)
13. M. Bando, K. Hasebe, A. Nakayama, A. Shibata, Y. Sugiyama, Dynamical model of traffic congestion and numerical simulation. Phys. Rev. E **51**(2), 1035–1042 (1995)
14. K. Hasebe, A. Nakayama, Y. Sugiyama, Dynamical model of a cooperative driving system for freeway traffic. Phys. Rev. E, Article no 026102, **68**(2), 6 (2003)
15. X.J. Wang, Z.L. Ning, L. Wang, Offloading in Internet of Vehicles: a fog-enabled real-time traffic management system. IEEE Trans. Indus. Inform. Article, **14**(10), 4568-4578 (2018)
16. A. Kesting, M. Treiber, D. Helbing, Enhanced intelligent driver model to access the impact of driving strategies on traffic capacity. Philos. Trans. Royal Soc. A: Math. Phys. Eng. Sci. **368**(1928), 4585–4605 (2010)

Chapter 5
Service Planning and Operation for Autonomous Valet Parking

Ziyi Hu, Jianyong Song, Yue Cao, and Yongdong Zhu

Abstract In recent years, the number of cars has been growing, and the demand for parking has surged. Meanwhile, with the progress of automobile industry and computer technology, the development of autonomous vehicle (AV) has further developed, resulting in the emergence of new parking service models. Users and the government urgently need new parking service models and parking resource management strategies to optimize parking behaviour. This chapter analyses several traditional parking models, introduces the new parking model—autonomous valet parking model especially long-range autonomous valet parking model in detail, and analyses its key issues. Further, this chapter introduces various parking resource management strategies, summarizes their advantages as well as disadvantages, and proposes future directions. This chapter also discusses the future research direction.

Keywords Parking service · Autonomous valet parking (AVP) · Long-range autonomous valet parking (LAVP) · Parking resource management

Abbreviations

AVs   Autonomous Vehicles
AVP   Autonomous Valet Parking

Z. Hu (✉) · J. Song · Y. Cao
School of Cyber Science and Engineering, Wuhan University, Wuhan, China
e-mail: ZiyiHu@whu.edu.cn

J. Song
e-mail: zaiyu404@qq.com

Y. Cao
e-mail: yue.cao@whu.edu.cn

Y. Zhu
Zhejiang Lab, Hangzhou, China
e-mail: zhuyd@zhejianglab.com

© The Author(s), under exclusive license to Springer Nature Singapore Pte Ltd. 2023
Y. Cao et al. (eds.), *Automated and Electric Vehicle: Design, Informatics and Sustainability*, Recent Advancements in Connected Autonomous Vehicle Technologies 3, https://doi.org/10.1007/978-981-19-5751-2_5

LAVP	Long-Range Autonomous Valet Parking
SAVP	Short-Range Autonomous Valet Parking
SC	Scheduling Centre
PL	Parking Lot
NP-Hard	Non-deterministic Polynomial-Hard

5.1 Introduction

The emergence of autonomous vehicles (AVs) has greatly reduced the driver's burden, with vision that everyone can facilitate the AV to travel (even if the user has no driver's license). However, this increases the traffic burden of the road. The congested road condition is not only caused by the increasing number of cars, but also due to improper parking resource management. According to work [1], 30% of traffic congestion in a typical downtown area is caused by the parking searching. Therefore, effective parking resource management measures are urgently needed.

Autonomous valet parking (AVP) is a new concept, which separates users from driving and parking. In a typical AVP workflow described in work [2], AV will automatically plan the optimal trip, pick up the user in the optimal pick-up spot, lead the user to the destination, and drive to the selected parking lot for parking after the user gets off the AV in the optimal drop-off spot. Users don't have to consider scheduling of travel planning or parking management for AV. It not only makes the user finish travelling and parking easier, but also considers the global traffic benefit maximization. AVP will be a new trend in the development of automatic parking service.

5.1.1 Motivation

When walking on the road, pedestrians usually complain about cars parked everywhere; When driving a car, users will say that parking slots are too difficult to find. With the reduction of automobile cost, there are more and more cars on the road, while the cars are parked in the parking slot 95% of the time according the survey result in work [3]. The road congestion and the "lack" of parking slots have bothered users. It seems to mean that the government urgently needs to build more parking lots. However, this solution sacrifices long-term benefits and only brings short-term improvement. More parking slots encourage more cars to travel, further causing traffic congestion. Providing convenient and cheap parking lots does not increase urban life or economic vitality. Ground parking lots will also affect the ecological environment. According to work [4], a better solution is to make full use of the existing parking resources including parking slots on the street and in the parking lot, called as parking resource management.

AVP service separates the driving phase from the parking phase. Users only need to give instructions, and AVs will automatically cooperate with the Scheduling Centre to receive decisions and complete operations related to driving and parking. As a key issue in parking services, parking resource management has played an important role and triggered many discussions. Parking resource management focuses on the management and allocation of parking slots, aiming to meet the needs of users and maximize the overall interests. The government or parking resource manager still needs more effective parking resource management strategies to improve the performance.

5.1.2 Main Contributions

Main contributions to this chapter are as follows:

(1) This chapter introduces the new parking service model AVP and its classification in detail. Among them, this chapter focuses on Long-Range AVP, including its system framework, workflow and key issues;
(2) This chapter elaborates on the key issue of parking services—parking resource management, states the existing problems and challenges. This chapter analyzes various parking resource management strategies, summarizes their advantages/disadvantages, and proposes suggestions for their further improvement.
(3) This chapter discusses the future development trend of parking service and related advanced technologies.

5.1.3 Structural Organization

The Sect. 5.2 introduces different parking service models, including traditional parking, smart parking and AVP. The Sect. 5.3 presents AVP in detail, including its system model and key issues. The Sect. 5.4 elaborates the parking resource management strategies. The Sect. 5.5 looks forward to the future development of parking service, and the Sect. 5.6 summarizes the whole chapter.

5.2 Background

5.2.1 Parking Service Model

5.2.1.1 Traditional Parking Model

In the traditional parking service model, users will not obtain additional information or advice except their subjective experience. They can only drive along the road to find an available roadside parking slot, or go to the parking lot and cruise in it to find an empty parking slot. According to the conclusion in work [5], it greatly increases the travel cost of users, deteriorates their parking experience, and further causes traffic congestion.

5.2.1.2 Smart Parking Model

With the development of communication technology, intelligent transportation system and Internet of things, the traditional parking model cannot meet users demand, thus smart parking model developing. Smart parking refers to the comprehensive collection, management and information release of urban parking slots through advanced technologies. It uses the sensing ability of sensor nodes to monitor and manage each parking slot, realizes the functions of parking slot management and provides users with parking slot guidance. In work [6], the smart parking model completes the unified planning and efficient management of parking resources. In work [7, 8], it also breaks the information island of traditional parking model by integrating urban offline parking slot information, which realizes the integrated service of real-time update, query, selection and navigation of parking slot resources. The smart parking model aims to maximize the utilization of parking slot resources, and it does.

5.2.1.3 Autonomous Valet Parking Model

With the development of AV, AVP shows incomparable advantages over other parking service models. The work [9] presents the several stages of AVP development. At first, AVP can only provide limited parking assistance, and users need to supervise the whole parking activity in the parking lot. With the development of wireless communication, users do not have to stay next to the AV. They can remotely execute and monitor the parking process through intelligent devices (such as mobile phones). Further, using path generation, accurate detection and other technologies, users can leave their AVs directly at the entrance of the parking lot through pre-training, and AV will drive to the designated parking slot automatically.

In a higher autonomy level of AVP, users do not have to go to the parking lot. They can get off at the designated place directly, and the AVs drive to the selected parking lot for parking. AVP separates the driving phase from the parking phase, so parking is no longer a problem for users. According to the urban structure, mobility and traffic scenarios, AVP can be divided into two sub-sections: short-range autonomous valet parking (SAVP) described in work [10] and long-range autonomous valet parking (LAVP) discussed in work [11]. This chapter will introduce AVP in detail in the Sect. 5.3.

5.2.2 Parking Resource Management

As mentioned earlier, the key to solve the parking problem is to shift from parking resource construction to parking resource management, which is also crucial to parking service. Parking resource management is responsible for information collection/dissemination of parking resources, as well as reasonably balancing user demands and allocating parking resources to users. The difficulties of parking management lie in:

(1) people generally own the traditional concept that parking resources should be free and available at any time;
(2) the responsibility of parking resource management is decentralized;
(3) parking resources are unbalanced and insufficient utilized;

How to apply appropriate parking resource management strategies to solve these problems is the focus of our discussion in the Sect. 5.4.

5.3 Autonomous Valet Parking

5.3.1 Savp

5.3.1.1 Overall

The development of computer vision, machine learning and autonomous car-manoeuvring technology promotes the development of SAVP, thereby enabling it to be realized in real scenes. In SAVP, AV must be trained in advance to complete autonomous parking. When initialized, the AV parks under the supervision to learn the new area and environment, that's to say, AV needs to be directed to the designated parking slot. After completing the learning process about the environment and trajectory under the guidance of the user, AV can fully realize autonomous parking.

Computer vision enables AV to scan and sense nearby objects, so as to perform necessary manipulation and navigation during parking. The latest SAVP system is also capable of cruising search of available parking slots in fully autonomous model. Due to the limitation of short-range effectiveness, SAVP works from the entrance of the parking lot.

5.3.1.2 SAVP Workflow

The user with parking demand will select a favourite parking lot. Then the user will leave the AV at the entrance of the parking lot, and the SAVP system will be responsible for parking the AV in the designated parking slot without the user's intervention. The user with travelling demand will wait for the AV to drive out of the parking lot. The role of SAVP is to move AV from parking slot to parking lot entrance and from parking lot entrance to parking slot.

5.3.1.3 Key Issues

(1) Parking Slot Detection

The purpose of parking slot detection is to obtain parking slot utilization information immediately. Here, using camera and image analysis technology for parking slot detection is a common method. Further, the application of machine learning and convolution neural network technology improves the performance of this solution, but it is limited by environmental conditions (such as light level). Another common method is the detection method based on ultrasonic sensor, because the cost of deploying ultrasonic sensor is low. However, it is also affected by sensor performance and its deployment location. Combining vision-based solutions with ultrasonic sensors can achieve better performance than using the single solution. Using deep learning to predict the parking slot state is also a novel solution.

(2) Path Planning

Path planning discovers the parking lot and plans the parking manoeuvre. It usually has two stages: first, it plans a feasible path from the starting point to the destination, considers the inherent kinematic constraints and converts the path into a path that the AV can follow. The goal of the first stage is generally to find a shortest path on the grid map. The search method is usually based on the shortest path algorithm proposed by Dijkstra. Second, it needs to study the local planning of vehicle moving trajectory, which needs to continuously control the angle and distance of vehicle moving forward or backward.

5.3.2 Lavp

5.3.2.1 Overall

Although SAVP solves the problems of difficult parking operations for users, it still does not completely "alleviate" users from parking. LAVP extends SAVP service, so that users do not need to worry about any decisions about parking. Users can get off at their favourite places, and then the AV can drive to the designated place or parking lot with the help of path planning and cooperative operation. Users with travelling demand only need to go to the boarding place determined by themselves, and then send instructions to the AV. AV can move from the parking slot to the boarding place in full autonomous model, so as to pick up the user.

5.3.2.2 LAVP Workflow

The Fig. 5.1 shows the overall architecture of LAVP. The Scheduling Centre plays a key role in LAVP. It receives the user's travel request and makes decision according to the user's requirements to determine the optimal Pick-up/Drop-off spot and the parking lot. The user interacts with AV at Pick-up/Drop-off spots. Users walk from the starting place to Pick-up spot and get on the AV. The AV carries users to the Drop-off spot, where users get off the AV and walk to the destination. In addition to sending a request to the Scheduling Centre and completing the boarding/alighting behaviour, users do not need to worry about anything. The driving behaviour is completed by AVs automatically, and the parking behaviour is completed under the interaction of AVs, parking lots and the Scheduling Centre.

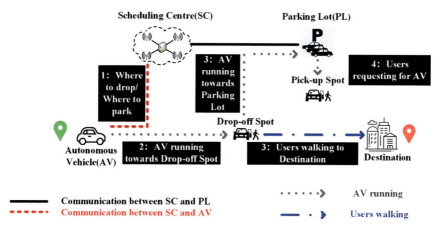

Fig. 5.1 LAVP system architecture

5.3.2.3 Key Issues

(1) Parking lots and Pick-up/Drop-off spots deployment

Pick-up/Drop-off spots connect AV and users in AVP. If the deployment of Pick-up/Drop-off spots is not optimal, it is bound to increase users' walking time, cause sidewalk congestion and affect traffic safety. The deployment of Pick-up/Drop-off spots needs to analyse the actual traffic conditions.

Pick-up/Drop-off spots are generally set up near commercial streets, entertainment places, large companies, industrial bases, schools, residential areas, etc. It should conform to a principle: Pick-up/Drop-off spots should be set up where there are more people going in/out, and the distance between Pick-up/Drop-off spots should change with the size of people flow. The locations of Pick-up/Drop-off spots shall enable users to get on/off AV safely and conveniently without affecting the normal driving of other AVs.

In AVP, parking lots are usually deployed at the edge of the city or in economically backward areas. This is because financially, the land utilization efficiency of parking slots is far lower than that of active buildings. Therefore, sections with low economic value should be selected for the construction of parking lots. The capacity of the parking lot is related to the parking demand of the area where it is located. However, due to the large mobility of AVs, the parking demand is uncertain. Before determining the capacity of the parking lot, it is necessary to conduct long-term analysis and short-term prediction of the traffic situation.

(2) Pick-up/Drop-off spots selection decision

When the Scheduling Centre selects Pick-up/Drop-off spots for users, two factors are usually considered: the driving cost of AV from the current location to Pick-up/Drop-off spots, and the walking cost of users walking from the current location to Pick-up/Drop-off spots (or walking from Pick-up/Drop-off spots to the destination). The driving cost can be expressed by driving time or energy consumed by AV, and the walking cost can be expressed by walking time or the distance. When calculating the total cost, the units of the two costs will be unified.

Generally speaking, the Scheduling Centre will select the Pick-up/Drop-off spots with the lowest total cost for users. However, some users may have their own preferences. For example, they prefer to walk more to reduce the energy consumption of AV. The Scheduling Centre can set different weight values when calculating the total cost according to users' different preferences.

(3) Sharing and interconnection of parking resources

Asymmetric information is a difficult problem to solve in the traditional parking model. Users often cannot obtain the latest parking resource information and finally make poor decisions. The sharing and interconnection of parking resources can effectively exchange information, which eliminates the problem of asymmetric information. In AVP, the Scheduling Centre monitors the utilization of all parking lots and the traffic situation near the parking lots. Similarly, it also receives AV parking requests and related information. The parking lot and AV are connected to each other through the Scheduling Centre to realize the sharing and interconnection of parking resources.

(4) Parking slot allocation control

The Scheduling Centre controls the parking slot allocation according to different parking resource management strategies. Like Pick-up/Drop-off spots selection decision, the Scheduling Centre often considers parking cost when deciding the optimal parking slot for users. Parking cost can be divided into three parts: the driving cost from the current location to the parking slot, the time cost of waiting for parking slot to be available and the parking fee cost paid for the parking slot.

Because the Scheduling Centre has global information, it can make the optimal decision for users. Optimal parking resource management strategies can improve the parking experience and traffic conditions, while inappropriate strategies will make users complain. The Scheduling Centre can fully allocate parking resources according to user requirements and the principle of First Come First Serve. The advantage of this strategy is to meet user requirements as much as possible, but it turns search parking into competitive parking, thus increasing the overall cost. Another strategy is to start from the whole and sacrifice part of the interests of individual users to maximize the overall interests. These two strategies accord with the first principle and the second principle of the famous definition of traffic network balance proposed by Wardrop in 1952. In the Sect. 5.4, this chapter will discuss these two and other parking resource management strategies in detail, analyse their advantages/disadvantages, and propose improvement suggestions.

5.4 Parking Resource Management Strategies

5.4.1 Users' Decision Making

When users need to travel from one place to another, they will choose the path with the lowest travel cost. The same is true when parking. Users will always choose the parking slot with the lowest cost for parking. In the traditional parking model, users need to choose parking slots according to their subjective feelings. But due to the problem of asymmetry information, users' own decisions are not necessarily optimal, resulting in additional costs. For example, without additional information, users tend to cruise around the destination first to find an available parking slot, which costs additional cruising time. However, the actual situation is that the user only needs to change places or wait beside an occupied parking slot for a short time to obtain a free parking slot. In AVP, the Scheduling Centre solves this problem by sharing and interconnecting global parking resources. The Scheduling Centre clearly knows when and where there are free parking slots and the cost of parking, and then allocates these free parking slots to users according to different strategies.

For parking resource management strategy that only considers the users' decision demands in work [12], the Scheduling Centre receives and processes the user's parking request according to the First Come First Serve principle. According to the current parking resources information of parking lots and the parking demand information of the users, the parking lot is selected for the users to minimize the cost of parking. After that, the Scheduling Centre updates the parking resources information. The Scheduling Centre will process the next request after the current request is processed.

The goal of the optimal decision problem is to allocate the parking lot with the lowest parking cost to each user. Each user must be allocated one parking lot and can only be allocated one parking lot. The work in [13] uses Exhaustive Attack method to solve this problem, and the sorting algorithm is used to further optimize the solution. It is of great significance in practice, for users have very strict requirements for the response time of the Scheduling Centre in reality.

This strategy conforms to the first principle of the definition of traffic network balance proposed by Wardrop: when the road users know the traffic state of the network and try to choose the shortest path, the network will reach the equilibrium state, that is, user equilibrium. The Scheduling Centre knows exactly the parking resource information and allocates the parking lot with the lowest parking cost to users, so as to finally achieve a balanced state.

The advantages of this strategy are: (1) It conforms to the natural cognition and behaviours of users. The parking resources obtained by each user must be the optimal under the current conditions, and the parking experience is improved; (2) Its solution is simple and works fast; (3) Due to the principle of First Come First Serve, fairness is guaranteed. However, it also has disadvantages: (1) The behaviour pattern of users has changed from search parking to competitive parking. In order to obtain better parking resources, users must park earlier, so as to increase the parking cost in a disguised form; (2) Individual optimization does not necessarily mean global optimization. A better way is to take advantage of the Scheduling Centre for global control to achieve better overall performance, which is discussed in the Sect. 5.4.2.

5.4.2 Global Controlling

In the first parking resource management strategy, the network will reach the user equilibrium state only when the user knows the parking resource information exactly and tries to select the parking resource with the lowest cost. However, in reality, the problem of asymmetry information will make users unable to obtain accurate resource information and difficult to achieve a balanced state. On the other hand, the user equilibrium state is not necessarily globally optimal. Furthermore, Wardrop proposed the second balance principle of traffic flow: all users follow the goal of "minimizing the total network cost" to choose the path. It is like a central manager coordinating the path selection behaviour of all users, and all users obey the command

5 Service Planning and Operation for Autonomous Valet Parking

of the central manager. As a result, the network state with the lowest total network cost is formed-system optimization.

In the system optimization state of the network discussed in work [14], the traffic network resources are optimally utilized and the traffic network benefits are brought into full play; The Scheduling Centre acts as the central manager, whose goal is to minimize the total user parking cost and finally achieve the system optimization state. In this parking resource management strategy, generally the system running time is discretized into shorter time periods, and the time nodes between time periods become decision points. The time period is generally set to be very short, so that the Scheduling Centre can respond to user requests in time. The Scheduling Centre collects the parking request information in the time period between the two decision points, updates the parking resource information, and allocates parking resources to all users requesting parking resources in this time period at the next decision point.

The allocation principle is that the total parking cost of these users is the lowest. The user unsatisfied with the parking resources allocated at this decision point can continue to request for the allocation of parking resources at the next decision point.

This parking resource management strategy can be modelled by mixed integer linear programming. The goal of the optimal decision problem is to allocate parking resources for each user at each decision point to minimize the total parking cost. The constraints include that users must be allocated and can only be allocated one parking resource, the limit of the number of parking resources, etc. The optimal solution of the mixed integer linear programming problem is the optimal allocation of parking resources for users.

The advantages of this strategy are: (1) The result must be globally optimal; (2) The problem modelling is simple and can be solved by commercial solvers; However, its disadvantages are obvious: (1) Although the modelling of mixed integer linear programming is simple, it is an NP-Hard problem. Even if the optimal solution can be obtained by commercial solvers, the scale of the problem it can solve is limited; In the actual scenario, the Scheduling Centre needs to respond to the parking requests of a large number of users in a very short time. When the scale of users is large, the solution time of the mixed integer linear programming problem becomes longer, which obviously does not meet the requirements of the actual scene; (2) It sacrifices the fairness of users. Later users may obtain a better parking resource, while the first users may not; This reduces the user's parking experience. Splitting to reduced the scale of the problem and adding constraints such as fairness can further improve the performance of this parking resource management strategy.

5.4.3 Parking Reserving

The work in [15] concludes that when parking resources are limited, the simple balanced traffic model (i.e. the two equilibrium states of user equilibrium and system optimization) is inefficient, because the competition for parking slots increases the parking cost of users by advancing the parking peak hours. Based on this conclusion, they further propose a parking permit mechanism. Some users will be allocated parking permits in advance, and the parking resources will be reserved for users with parking permits. Parking permit can be regarded as a simple reservation mechanism, through which parking resources are reserved for users in advance without competition.

In fact, the first two parking resource management strategies this chapter discussed in the Sect. 5.4.1 and the Sect. 5.4.2 are also based on the parking reservation service mechanism, because users will obtain available parking resources from the Scheduling Centre before arriving at the destination. It is equivalent to that the Scheduling Centre allocates and retains parking resources for users in advance. The parking reservation service mechanism effectively solves the problem of "multiple cars chasing the same spot", and provides guaranteed parking resource reservation for user.

According to the work [16], the workflow of the parking reservation mechanism is divided into four steps shown in Fig. 5.2:

- The Scheduling Centre monitors and aggregates the parking resource information of each PL. Users send parking reservation requests and related information to the Scheduling Centre;

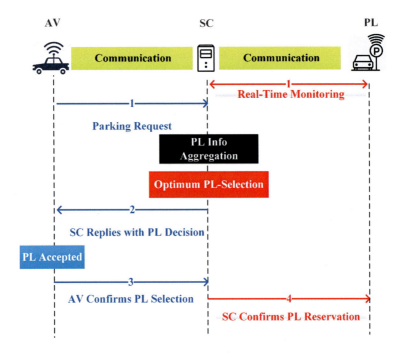

Fig. 5.2 The workflow of parking reservation mechanism

- The Scheduling Centre pre-allocates the parking resources in the optimum PL for users according to the current parking resource information, request information and parking resource management strategy (like based on user equilibrium or system optimization), then sends the pre-allocation results to users;
- The user satisfied with the pre-allocated parking resources will send a confirmation message to the Scheduling Centre, otherwise the user will require the Scheduling Centre to re-allocate the parking resource;
- The Scheduling Centre receives the user's feedback. If the user accepts, the parking resource will be reserved for the requested user to ensure that only the user can obtain it; Otherwise, the Scheduling Centre will re-allocate the parking resource for the user.

The Scheduling Centre needs to ensure that users can only obtain the reserved parking resources within the reserved time. Once the reservation time is exceeded, the reservation will be voided and the parking resources will be reassigned to other users. The parking lot and the Scheduling Centre need to jointly ensure the matching between users and reserved parking resources, and the reserved parking resources cannot be used by other users.

The advantages of parking reservation service mechanism are: (1) It reduces the competitive parking cost caused by limited parking resources such as cruising time; (2) It enables the Scheduling Centre to obtain traffic information in advance, and then take more appropriate control strategies according to the current/future conditions, so as to effectively reduce traffic congestion. Its disadvantage is that users who do not participate in the reservation service will suffer greater losses: when they arrive at their destination, they may find that the available parking resources have been reserved. Even if there are free parking slots at that time, they cannot obtain them. A better approach is that the parking lot only provides part of the parking slots for advance reservation, and the remaining parking slots are provided to users for free competition.

5.4.4 Slot Sharing

The progress of the Internet and online payment has led to the vigorous development of the sharing economy. Sharing bicycles/Portable chargers are successful cases. The essence of sharing economy is to transfer redundant ownership to others and let them obtain temporary utilization rights, so as to create value for both the supplier and the demander. Shared parking conforms to the essence of sharing economy. The core of shared parking is that units or individuals with dedicated parking resources rent idle parking resources to other users who need them.

In real life, for users with personal parking slots in the community, they usually do not utilize their own parking slots during working hours, but park their cars near the work area. At this time, their parking slots are idle. Sharing parking will rent their idle parking resources to others, thus bringing benefits.

The shared parking requires a third-party market platform through which users can rent or exchange idle parking resources. As a link between the supplier and the demander, the third-party platform enables the supplier and the demander to trade through it through the establishment of a series of mechanisms, such as mobile LBS application, dynamic matching algorithm and mutual evaluation system. The third-party platform collects supply/demand information, thus realizing the optimal match between suppliers and demanders to meet their demands. In work [1], stable matching algorithm and price compatible transaction cycle/chain mechanism are used to achieve this goal.

Shared parking is the combination of shared thinking and big data application, which is in line with the essence of shared economy. Its advantages are: (1) It realizes the reuse of parking resources, and effectively utilizes the existing parking infrastructure; (2) It can bring benefits to both users and third-party platforms, as well as driving the commercial and economic development around the parking lot.

However, it is still difficult to implement in reality: (1) The management of shared parking is difficult and the cost are large. Many users' idle parking resources are located in the parking lot of the community, and the estate is responsible for the community management. It is difficult for the third-party platform to ensure the impact of shared parking on the security of the community; (2) Shared parking needs to coordinate the interests of multiple parties, such as the supplier, the demander, the estate and the third-party platform. Factors such as low profit, low default cost and difficult negotiation will reduce the enthusiasm of the supplier and the demander, and it is difficult to realize sharing in the end. The real realization of shared parking needs the guidance of the government and the participation of enterprises. The government needs to formulate overall urban solutions, guide enterprises to build large parking platforms, promote the release of parking legislation and policies, and formulate relevant technical standards.

5.4.5 Dynamic Pricing

Parking resource is a kind of commodity provided to users with cars, so it also abides by the supply–demand relationship of commodities. As mentioned earlier, providing free parking slots is not the wisest parking resource management strategy. For example, if all parking resources are free, the parking lot which is closer to the destination will attract more users, while the farther parking lot will be ignored, resulting in the uneven utilization rate of each parking lot. Free parking resources provide economic power for users to cruise around the destination.

The core idea of dynamic pricing is to take advantage of different parking resource prices to balance the utilization of parking resources in each region. The pricing for parking resources in areas where the current utilization rate exceeds the expected utilization rate will be increased, and the pricing for parking resources in areas with opposite conditions will be reduced. Finally, the utilization rate of parking resources in each area will be close to the expected utilization rate. Expected utilization rate is a fixed value set in advance. In work [17], 85% is considered a suitable value. Pricing adjustment needs to consider larger price changes in the short term or light price adjustments in the long term.

In 2008, the San Francisco Municipal Transportation Agency planned the SFpark project [18]. The strategy of SFpark project is shown in Fig. 5.3. The SC of SFpark project monitors the utilization of parking resources in each block, and calculates the average utilization rate. After a specific time period, SC will compare the average utilization rate with the target rate (60–80%). The block with average utilization rate of parking resources below/above the target rate will obtain a parking price adjustment by SC. After that, SC will continuously monitor the impact of current prices on parking resource utilization and dynamically adjust parking prices. The project had been implemented since July 2011, which significantly improved the availability of parking slots, reduced vehicle greenhouse gas emissions, and improved the economic competitiveness of the region.

The dynamic pricing brings following benefits: (1) Its implementation cost is low and can improve the actual income generated by parking resources; (2) It has no privacy or security issues; (3) It uses the existing parking management infrastructure; Its disadvantage is that the interval between price updates is too long, which limits the ability of the model to deal with anything other than the average parking demand, and it is difficult to respond to special events. Optimizing the operation cycle to meet the changing demand, and adjusting the price according to the forecast of future

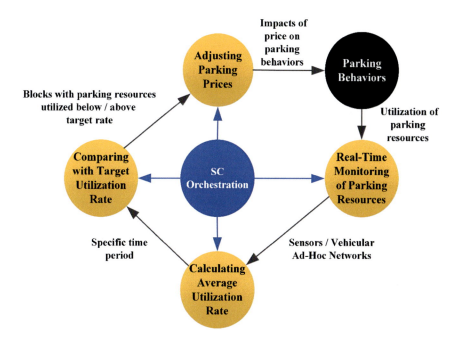

Fig. 5.3 The strategy of SFpark project

demand rather than the past utilization data will better improve the performance of dynamic pricing.

5.5 Future Directions

5.5.1 Blockchain Technology

Blockchain is a distributed shared ledger/database, which has the characteristics of decentralization, non-tampering, traceability, collective maintenance, openness, transparency and so on. These characteristics ensure the "honesty" and "transparency" of the blockchain, solve the problem of asymmetry information. It realizes the cooperation, trust and concerted action among multiple subjects. In work [19], the application of blockchain in parking service realizes the comprehensive transmission of parking resource information in the whole network, verifies the accuracy of information and improves the security of parking service. However, because the blockchain can only allow the virtual currency bitcoin, and the speed of generating a block is slow, it still needs more research and practice to apply the blockchain technology to the parking service in practice.

5.5.2 Deep Reinforcement Learning

Deep learning is a representation learning method of data in machine learning. Reinforcement learning obtains an optimal strategy by exploring the unknown environment, establishing the environment model and learning. Deep reinforcement learning combines the perception ability of deep learning with the decision-making ability of reinforcement learning, which can be controlled directly according to the input information. In work [20], deep reinforcement learning method is effectively used to decide how to better allocate taxi drivers to users in taxi-hailing apps. Most of the current parking resource management strategy can only ensure the optimal profit in the current or a short future time slot, while deep reinforcement learning is able to provide long-term solutions. However, it faces the same problem as blockchain technology, that is to say, it is difficult to implement in actual scenarios.

5.6 Conclusion

In this chapter, three parking service models are discussed with a focus on AVP (including both SAVP and LAVP). AVP is also the most digitalized parking service model, which effectively solves most of the problems encountered by users in parking. This chapter also introduces five different parking resource management strategies, analyses their working processes, and proposes a series of suggestions on parking service improvements.

References

1. D. Shoup, Cruising for parking. Trans. Policy, **13**(6), 479–486 (2006)
2. M. Chirca, R. Chapuis, R. Lenain, Autonomous valet parking system architecture, in *2015 IEEE 18th International Conference on Intelligent Transportation Systems* (2015), pp. 2619–2624
3. British National Travel Survey, https://assets.publishing.service.gov.uk. Accessed April 2018
4. D. Shoup, *The High Cost of Free Parking* (APA Planner, Chicago, IL, USA, 2005)
5. K.W. Axhausen, J.W. Polak, M. Boltze, Effectiveness of parking guidance and information systems: recent evidence from Nottingham and Frankfurt am Main. Traff. Eng. Control **35**(5), 304–309 (1993)
6. T. Lin, H. Rivano, F. Le Mouël, A survey of smart parking solutions. IEEE Trans. Intell. Transp. Syst. **18**(12), 3229–3253 (2017)
7. J. Lin, S. Chen, C. Chang, G. Chen, SPA: smart parking algorithm based on driver behaviour and parking traffic predictions. IEEE Access **7**, 34275–34288 (2019)
8. A.O. Kotb, Y. Shen, Y. Huang, Smart parking guidance, monitoring and reservations: a review. IEEE Intell. Transp. Syst. Mag. **9**(2), 6–16 (2017)
9. M. Khalid, Y. Cao, N. Aslam, C. Suthaputchakun, M. Arshad, W. Khalid, Optimized pricing & scheduling model for long range autonomous valet parking. Int. Conf. Front. Inform. Technol. (FIT) **2018**, 65–70 (2018)
10. U. Schwesinger et al., Automated valet parking and charging for e-mobility. IEEE Intell. Veh. Symp. (IV) **2016**, 157–164 (2016)
11. M. Khalid et al., Towards autonomy: cost-effective scheduling for long-range autonomous valet parking (LAVP). IEEE Wirel. Commun. Network. Conf. (WCNC) **2018**, 1–6 (2018)
12. X. Zhang, F. Yuan, Y. Cao and S. Liu, Reservation enhanced autonomous valet parking concerning practicality issues. IEEE Syst. J. **16**(1), 351–361 (2022)
13. C. Tang, X. Wei, C. Zhu, W. Chen, J.J.P.C. Rodrigues, Towards smart parking based on fog computing. IEEE Access **6**, 70172–70185 (2018)
14. Y. Geng, C.G. Cassandras, New "Smart Parking" system based on resource allocation and reservations. IEEE Trans. Intell. Transp. Syst. **14**(3), 1129–1139 (2013)
15. X. Zhang, Y. Hai, H.J. Huang, Improving travel efficiency by parking permits distribution and trading. Transp. Res. Part B Methodol. **45**(7), 1018–1034 (2011)
16. A.O. Kotb, Y. Shen, X. Zhu, Y. Huang, iParker—a new smart car-parking system based on dynamic resource allocation and pricing. IEEE Trans. Intell. Transp. Syst. **17**(9), 2637–2647 (2016)
17. X.T.R. Kong, S.X. Xu, M. Cheng, G.Q. Huang, IoT-enabled parking space sharing and allocation mechanisms. IEEE Trans. Autom. Sci. Eng. **15**(4), 1654–1664 (2018)
18. D. Shoup, Q. Yuan, *Parking and City* (China Architecture & Building, China, 2020)
19. SFpark, http://sfpark.org
20. L. Wang, X. Lin, E. Zima, C. Ma, Towards Airbnb-like privacy-enhanced private parking spot sharing based on blockchain. IEEE Trans. Veh. Technol. **69**(3), 2411–2423 (2020)
21. B.Y. Cai, R. Alvarez, M. Sit, F. Duarte, C. Ratti, Deep learning-based video system for accurate and real-time parking measurement. IEEE Internet Things J. **6**(5), 7693–7701 (2019)

Chapter 6
Navigation Service Optimization for Electric Vehicle

Xinyu Li, Yue Cao, and Shuohan Liu

Abstract The development of electric vehicles (EVs) has reduced environmental pollution and brought many conveniences to urban life, driven by advances in sustainable energy development. Benefited from the above advantages, EVs are considered to be the representative of the next generation of transportation electrification and gain popularity among different countries and regions. However, range anxiety has still been a significant problem restricting the development of EVs. Therefore, navigation service optimization is necessary for the long-term prosperity of the EVs market. In this chapter, we attempt to introduce some related work and new technologies in this domain. The chapter demonstrates the development of EVs, different EV charging models and the navigation service system. Then, we introduce the navigation service optimization from plug-in and battery swapping models respectively. Finally, we show some burgeoning charging service methods.

Keywords Electric vehicles · Energy supplement · Charging service optimization · Transportation planning · E-mobility

Abbreviations

ICE   Internal Combustion Engine
EV   Electric Vehicle
BEV   Battery Electric Vehicle
PHEV   Plug-in Hybrid Electric Vehicle

X. Li (✉) · Y. Cao
School of Cyber Science and Engineering, Wuhan University, Wuhan, China
e-mail: 645888651@qq.com

Y. Cao
e-mail: yue.cao@whu.edu.cn

S. Liu
School of Computing and Communications, Lancaster University, Lancaster, UK
e-mail: s.liu37@lancaster.ac.uk

© The Author(s), under exclusive license to Springer Nature Singapore Pte Ltd. 2023
Y. Cao et al. (eds.), *Automated and Electric Vehicle: Design, Informatics and Sustainability*, Recent Advancements in Connected Autonomous Vehicle Technologies 3, https://doi.org/10.1007/978-981-19-5751-2_6

HEV	Hybrid Electric Vehicle
CS	Charging Station
BSS	Battery Swapping Station
SOC	State of Charge
IoV	Internet of Vehicles
ITS	Intelligent Transportation Systems
VANET	Vehicular Ad-Hoc Network
MANET	Mobile Ad-Hoc Networks
DTN	Delay-Tolerant Networks
V2V	Vehicle-to-Vehicle
V2L	Vehicle-to-Load
V2X	Vehicle-to-Everything
V2H/B	Vehicle-to-Home/Buildings
V2I	Vehicle-to-Infrastructure
GA	Global Aggregator
TOU	Time-Of-Use
RES	Renewable Energy Source
SCUC	Security-Constrained Unit Commitment
RT-SLM	Real-time Smart Load Management
CSP	Closest Station Policy
MWT	Min Waiting Time Policy
PSCC	Power System Control Center
ILP	Integer Linear Programming
MILP	Mixed Integer Linear Programming
RO	Robust Optimization
MIQP	Mixed Integer Quadratic Programming
MEC	Mobile Edge Computing
LMO	Lithium Manganese Oxide
NCA	Nickel-Cobalt-Aluminum
MPU	Mobile Power Unit

6.1 Introduction

Recent studies about climate change are mandating a drastic reduction of air pollution from CO_2 emissions. Traditional internal combustion engine (ICE) vehicles inevitably bring a lot of environmental issues, such as air pollution, energy waste, and so on. During the last decade, the increased utilization of electric vehicles (EVs) has been at the center of much attention among all solutions for improving the sustainability of transportation systems. Many countries have proposed positive policies and regulations to encourage the development of EVs. It has been widely regarded as a promising technological measure to deal with environmental issues.

However, different from traditional gasoline-powered vehicles, range anxiety and long charging time is the main drawback for the rapid development of EVs. At present, the limited cruising range of EVs is mainly concentrated in 100–200 km and frequent recharging is therefore required. The expected waiting time at a station is usually unavailable for EV drivers before they arrive at it due to lacking live information about charging stations. Furthermore, an uncoordinated charging plan will lead to the number of EVs driving to some specific hotspot charging station, which results in a long waiting time and increases people's dissatisfaction with EVs. Meanwhile, a large-scale EV charging without carefully planning will inevitably cause heavy pressure on the grid. This also results in a terrible service experience for EV drivers considering the load pressure of the grid. The reason is that the charging power will be adjusted to ensure the stability of the grid especially during the peak period, which increases the charging waiting time and charging service rejection rate.

To address the above problems standing in the way of the booming use of EVs, it has been demonstrated that charging scheduling is an effective scheme for navigation service optimization. In this chapter, the navigation service optimization process including the temporal and spatial domain is introduced. Besides, various energy supplement modes for EVs are also introduced. Moreover, the research methods of optimization solver based on the above various charging modes are also discussed.

6.2 Background

6.2.1 Type of EVs

Battery EVs (BEVs), plug-in hybrid EVs (PHEVs), and hybrid EVs (HEVs) are the main types of EVs in the transportation system. On one hand, a rechargeable battery without an internal combustion engine is adopted for BEVs. On the other hand, a battery and a gasoline engine both exist in the PHEVs. During the travel, a rechargeable battery is used before turning on the gasoline engine by the PHEVs. Different from the above two types of EVs, the HEVs contain an electric motor, a battery system, a gasoline engine, and a fuel tank. For the main component in the HEVs, firstly, the gasoline engine and electric motor always start simultaneously. Then, the load is transferred to the electric motor to actuate an EV. It should be noticed that any external sources are not needed to charge the battery. The HEVs are designed with the internal regenerative braking process to recharge the battery. Some classical EVs are appeared on the market and earn a good reputation, such as Toyota Prius Hybrid, Honda Civic Hybrid, and Toyota Camry Hybrid.

6.2.2 EV Charging Use Case

The research on EV charging service optimization has mainly been divided into two cases: **Parking Mode** and **On-the-move Mode**.

6.2.2.1 Parking Mode

More recently, there has been extensive research on charging scheduling about when/whether to charge (parking mode), which is regarded as temporal domain research. In this scenario, EVs are usually parked at home, charging stations (CSs), or parking lots. The vehicle is plugged into a charging plot at a parking spot and a sign-on order is controlled by the global aggregator. EVs transmit self-information, such as departure time, the current state of charge (SOC), desired SOC and so on to the global aggregator through ZigBee, Wi-Fi and cellular network. The global aggregator collects charging information from EV-side and real-time electrical load from the grid-side to decide the charging period for EVs. The above discussed temporal charging scheduling process is modeled as an optimization problem subject to some specific constraint conditions. For the optimization problem, a collection of charging required EVs, grid loads, and global aggregator information are regarded as inputs. Accordingly, outputs of the optimization problem are charging plans for the parked EVs set (when each EV begins and finishes charging).

6.2.2.2 On-the-Move Mode

As compared to massive literature on the traditional when/whether to charge problem, studies on where to charge (on-the-move mode) represent a growing field for charging scheduling and the problem has not been fully investigated. In this case, EVs are assisted to drive toward a CS according to some approximate criteria when the charging demand emerged. Notice that the spatial charging scheduling problem, namely the CS-selection process can be regarded as a navigation service optimization problem for EVs. The navigation service for EVs are usually performed with the following metrics:

- **Average waiting time**: The average time duration for an EV to spend at the selected CS, including the waiting time for charging and the charging duration.
- **Travel distance**: The distance for an EV to travel at the selected CS.
- **Economic factor**: The money the EV spends at the CS and CS purchases electricity from the grid.

The suitable CS for EVs brings a satisfied charging experience, mitigating range anxiety, balancing different entities, and improving driver's confidence for EVs development. Therefore, it can be concluded that the navigation service is of great importance for EVs to make the best CS-selection decision.

6.3 Navigation Service System

The Navigation Service System provides the basic framework for the Navigation Service. We will introduce the communication, main entities, and system cycle of the Navigation Service in this section.

6.3.1 Network Communication for Navigation Service

It has been widely accepted that booming communication technologies involved in the Internet of Vehicles (IoV) help to improve charging scheduling service, increase user satisfaction and inspire some emerging domains, such as Intelligent Transportation Systems (ITS), smart city, smart grids, and so on. A classic two-layer architecture with the service stratum and the communication infrastructure or transport stratum were usually regarded as the realization of the above-discussed domains [1, 2]. To be specific, the service stratum consists of different service entities. Furthermore, these services are able to be organized and a specific application is implemented with this method. Besides, the communication between different service entities and clients is achieved by the network infrastructure.

In term of ITS, which is our major concern, vehicular networks (VANET) has played an important role in implementing ITS, due to its brilliant performance on a number of applications, such as on-board Internet access, road safety and so on [1]. On the service stratum side, VANET consists of an application layer with services supply, such as pushing travel information to EV drivers. On the transport stratum side, VANETs is the embodiment of mobile ad-hoc networks (MANET) in the vehicular world. Notice that VANETs are usually implemented with wireless network protocols. But classical wireless network protocols usually may not match with the high mobility property of real-world vehicles. Recently, delay-tolerant networking (DTN)-based communications technologies are developed for low latency connection, faster data transfers, and safer and more reliable message estimation. Besides, some neoteric mobile communications technologies, such as WiFi, Bluetooth, 4G, and even 5G networks are also applied on Vehicle-to-Vehicle (V2V) and Vehicle-to-Infrastructure (V2I) communications.

6.3.2 Network Entities

In most of the existed works, a centralized controller, e.g. Global Aggregator (GA) usually plays an important role in navigation service. Three major network entities for the service can be concluded as follows:

- **Electric Vehicle (EV)**: For each EV, it has state information, named Status Of Charge (SOC). The SOC threshold indicates the EV state value needed charging

service. When the current SOC is below the setting SOC threshold, the EV with charging requirement will communicate with the GA, and the most suitable CS is recommended.
- **Charging Station (CS)**: The EVs complete the charging operation at CS, which is located at a designed coordinate point over the city map. There are multiple charging slots in each CS and the local condition including the number of available charging slots, charging EVs are monitored by the GA.
- **Global Aggregator (GA)**: The CA monitors the CS condition state and EV charging requirement in a centralized manner. EVs with charging requirements sent information to CA and CA guide the EV to the most appropriate CS with collected the CSs' condition information.

6.3.3 System Cycle

Figure 6.1 depicts the four main phases for EV charging navigation service:

- **Driving Phase**: The EV is moving on the road and driving toward its trip destination.
- **Where to Charge Phase**: When the current SOC is below the setting SOC threshold, the EV will find a CS to recharge its battery. The charging requirement and information from CSs are collected by the decision-maker, then the decision on where to charge is sent to EV drivers.
- **Waiting Charging**: When the EV arrives at the recommended CS, the EV will immediately charge its battery if there is an available charging plot in CS. Otherwise, the EV will wait at the CS until an EV finish charging.

Fig. 6.1 EV charging navigation service cycle

- **Battery Charging Phase**: The EV is refueled to the required SOC via the plug-in charger at CS. After the battery charging phase, the EV will leave the CS and turn to the driving phase.

6.4 Navigation Service Optimization Based on Plug-In Charging Mode

6.4.1 Introduction for Plug-In Charging Mode

During the development of EVs in the last decade, PHEVs and BEVs have taken the majority part of the EV market. Different from HEVs with two power sources, PHEVs and BEVs are only powered by batteries. Therefore, it is a crucial issue for PHEVs and BEVs to refill energy when their batteries are depleted. The plug-in charging mode is the most common way for EVs to complete the power replenishment. To achieve the recharging operation, the EV drives to the selected CS and is connected to the grid with the plug-in charger.

With the booming of a large scale of PHEVs and BEVs, existing grid system and the CS operator may be heavily impacted. A surge of charging demands inevitably causes huge instability to the fragile power distribution systems, which eventually effects the QoE of charging service for EV drivers. Considering these factors, optimal charging scheduling is necessary for EVs' navigation service. In this section, we will briefly introduce the navigation service optimization about plug-in charging strategies.

6.4.2 Optimal Plug-In Charging Strategy

As discussed above, there are many factors are concerned with the service optimization. The charging optimization strategies applied to EVs can be summarized as follow:

- **Instant Charging Strategy**: Once the EV is connected with the grid through the plug-in charger, the charging management system will provide service with the full power to the EV immediately. This is a common charging scenario happened at home. In this scenario, the EV is back to home and charged in the evening. Then, the EV is driven to the workplace at the morning after charging the battery from the initial SOC to the full capability. This is also regarded as an uncontrolled strategy without considering additional factors.
- **Predetermined Start Time Charging Strategy**: Different from the above instant charging strategy, the EV is not immediately charged although it is connected to the grid. It starts to be charged after a predetermined time, such as 10:00 pm. In this

way, the minimal charging cost can be obtained for EVs considering Time-Of-Use (TOU) electric price. This is also named as delayed charging strategy.

- **Automatic Off-Peak Charging Strategy**: The above predetermined start time charging strategy considers the charging cost for EVs but ignores the peak-valley characteristic of the grid. In this scheme, CSs are cooperated with the utility company and obtained the peak-valley period. Then, the charging power and the service acceptance rate are both adjusted to ensure the stability of the grid.
- **Optimal Charging Strategy**: In this scheme, the real-time traffic flow data and the state of the electric grid are obtained. The optimal charging target and charging scheduling strategy are derived with the constraint of the above real-time information. Finally, the set optimal target such as: minimizing charging cost and service waiting time, maximizing social welfare can be completed. This is also called a smart and efficient charging scheduling strategy, and majority of literature focuses on it.

6.4.3 Spatial and Temporal Characteristic Under Plug-In Charging Mode

6.4.3.1 Temporal Dimension

In the temporal domain scenario, EVs are parked without considering the property of mobility, and they are regarded as the static load connected to the grid. Researches of service optimization on temporal dimension usually focus on "when/whether to charge". To minimize the impact of large-scale EVs charging and provide the optimal navigation service, the plug-in charging scheduling strategies should be paid special attention. The service capability of the CS, the electric load of the grid and the charging cost of EVs are the main factors taken into account in related works.

There are different focuses for different participants in the temporal charging scheduling optimization problem. To be specific, some works indicate the operator cares about maximizing financial profits and the supply/demand balance (peak shaving and valley filling) on **grid-side**. Renewable energy sources (RESs) with wind and hydropower were considered in [3]. In this literature, a stochastic security-constrained unit commitment (Stochastic SCUC) model is used to analyze the behavior of PEV sets and RESs in power systems to minimize the cost of power supply. Li et al. [4] considered the PEV fleet's impacts on the power system and develop a distributed PEV charging algorithm to mitigate peak load, so that peak shaving and valley filling can be achieved. With respect to coordinated PHEV charging, Sortomme et al. [5] proposed three optimal charging algorithms to minimize the loss in the distribution system.

Some other works have tried to study the optimization problem from **EV-side** such as maximizing the experience of driver charging service, maximizing charging energy for EV fleets, and minimizing charging cost. Wen et al. [6] considered the distribution power system demand constraints and maximize user convenience function. A convex relaxation optimization method is proposed to solve the PEV-charging scheduling problem. He et al. [7] presented a globally and locally optimal scheduling scheme for G2V (grid to vehicle) and V2G (vehicle to grid) situations. The strategies are proved to minimize the total cost for drivers and handle a large number of EVs with random arrivals. Su et al. [8] used the estimation of distribution algorithm (EDA) to optimally managing a large number of PHEVs at a parking station, which aims to maximize the average SOC of EVs.

Accordingly, some works have focused on **both grid-side and EV-side**. Considering energy supply cost with real-time price and drivers' preferred charging periods, Deilami et al. [9] proposed a real-time smart load management (RT-SLM) control strategy for PEVs fleets' coordinated charging. Kim et al. [10] integrated the various driver charging requirement constraints and peak power supply into the efficient charging schedule strategy. Jin et al. [11] used a cyber-physical system approach to describe a large number of EV fleets charging behavior. A decentralized scheduling algorithm is proposed in the paper to classify EVs into different groups and each group is scheduling with group demand.

Some literature have been derived from the **CS-side** including charging station or parking station that provides charging service. Yao et al. [12] studied demand response, which is an important concept in smart grid considering the balance between demand and supply. This paper is formulated with the objective function of maximizing the number of serviced EVs and minimizing the expenditure cost in order to ensure the profits of the parking station. Emmanouil et al. [13] proposed a decentralized and dynamic method to maximize charging stations' profit and minimize the undesirable impact. The scheme has been proved to achieve similar congestion management effect with centralized solution, but it exhibits robust to large-scale EVs. Wang et al. [14] proposed a joint admission and pricing operation scheme for CSs. With jointly considering the acceptance rate of EV charging service and pricing, a charging optimization scheme is presented to maximize the profits of the CSs. Apart from the literature discussed above, Mukherjee et al. [15] also introduced service optimization methods in the temporal domain scheduling problem in detail.

6.4.3.2 Spatial Dimension

The above subsection introduces the development of navigation service optimization from the perspective of different participants based on the temporal dimension, which aims to solve the problem "whether/when to charge". The research on temporal dimension is certainly an important part for the navigation service optimization. However, these works are proposed with the assumption that EVs are static load without considering the characteristic of mobility. Considering this point, some literature devotes to solving "where to charge", namely CS-selection.

Gharbaoui et al. [16] developed a smart management system for Electric Vehicle (SMS-EV) based on an ICT infrastructure to support the CS-selection process. Furthermore, two selection policies, one is Closest Station Policy (CSP), and the other is Min Waiting Time Policy (MWT) are studied. The policies are compared based on the indicator of average waiting time. The MWT shows remarkably superior performance especially at urban area in comparison with the CSP by the system evaluation. This paper is an earlier study on CS-selection and considering the difference between minimum waiting time and closet station policy. Lee et al. [17] considered the CS-selection under a specific tour-and-charging scheduler scenario. The proposed 3 courses are compared with the traveling salesman problem solver, providing the effectiveness of proposed courses in reducing the waiting time. For EV taxis drivers, the charging waiting is a no-negligible portion to the total work hours and heavily affects their daily revenue. Therefore, Tian et al. [18] designed a system for real-time charging station recommendation with GPS data. Combining EV's historical charging behavior and real-time GPS data, the system predicates the EV's next statute and recommends it to the CS with minimal charging waiting time. Guo et al. [19] jointly considered the traffic conditions and the status of the power grid. Then, an intelligent transport system (ITS) is studied including a power system control center (PSCC), an ITS center, CS, and EV. The EV calculates the minimal total time for charging with data transferred among these modules.

6.4.4 Navigation Service Optimization in Different Manner Under Plug-In Charging Mode

6.4.4.1 Centralized Navigation Service Optimization

The centralized navigation service optimization is the most common service optimization scheme because it is able to access the global information and accordingly obtain the most optimal solution with a suitable model. The GA plays an important role in centralized navigation service. The centralized entity manages all charging demand across the whole network and recommends the suitable CS for EV drivers according to the status of CSs and the traffic flow on the road. We should notice that some researches have also called the CA as the city brain or the recommendation system and so on. Anyway, a centralized entity for global information aggregated, processing and the optimal decision is necessary in centralized navigation service optimization.

Timpner et al. [20] designed a V-Charge project, which provides a service system combining autonomous valet parking with e-mobility based on a central server. The scheme manages and assigns CSs to EVs with the aggregated information about charging and parking resources. Simulations verify the effective utilization of charging resources and increase the satisfaction of EVs with the practical data from Hamburg Airport and the City of Braunschweig, Germany. Emmanouil et al. [14] compared the centralized solution and the distributed solution in EV charging scheduling problem regarding to the QoS of drivers and the profit of CSs. It is proved that the centralized solution indeed achieves the optimal performance than the distributed solution, especially considering small-scale EVs. Although it still exhibits desirable performance on large-scale EV operations, the enormous calculation cost is inevitable. Wei et al. [21] developed an intelligent charging management system to control the charging behavior of EVs at parking garage for the purpose of maximizing the benefits for the EV drives and the charging operator. The charging management system, as a central controller, executes an admission control mechanism to guarantee the QoS of each EV. An adaptive utility oriented scheduling algorithm is accordingly proposed to optimize the utility function from the charging operator perspective. Simulations based on the practical battery characteristic and Time-Of-Use (TOU) electric price exhibit the effectiveness of the proposed algorithm. Cao et al. [22] utilized the global aggregator to achieve CS-selection decision based on reservation. Moreover, the influence of traffic condition has also been studied and a charging strategy is proposed to minimize drivers' trip duration. Simulations under the Helsinki city scenario is performed to present the desirable performance of the proposed scheme considering EVs' trip duration and the various metrics.

6.4.4.2 Distributed Navigation Service Optimization

As discussed above, the centralized navigation service optimization scheme is able to obtain the optimal solution and be implemented at an acceptable computation cost, especially for small-scale EV operations. However, the centralized method usually suffers from the problem of computation complexity, which usually needs a large amount of calculation resources and time cost to deal with the large-scale agents. Besides, all information is aggregated to the central entity and inevitably causes heavy load on it. If the failure of the central entity occurs, the whole system will face a huge challenge. Because all entities participating in charging send information containing its key feature to the central entity, the privacy issue is also an important point that should be focused on. Considering these disadvantages, the distributed scheme has been proposed as an alternative.

Emmanouil et al. [14] also proposed the distributed solution apart from the centralized scheme regarding the charging problem discussed in the last subsection. Simulations based on the road network of a UK city are performed, which proves the centralized scheme achieves the most desirable performance while not robust to large-scale. The distributed algorithm can be implemented with thousands of agents at cost of little performance degradation. Alsabbagh et al. [23] proposed a distributed charging management mechanism for EVs according to different customer behaviors. The paper decomposes the charging problem of EVs into two sub-problems, namely the pricing game and the power distribution game. The first sub-problem focuses on the charging admission charging control with the different driver behaviors to charging price. The second sub-problem aims to solve how to allocate the charging power when EVs connected into the grid. Both sub-problems are solved iteratively in a distributed way. Yang et al. [24] considered the problem of coordinate EV charging with locally generated wind power of buildings. An EV-based decentralized charging algorithm is proposed to overcome the necessity of global information and computing complexity with the increase of EV density. The proposed algorithm is also proved to have similar performance with the centralized solution.

6.4.5 Optimization Solving Methods for Navigation Service

The Navigation service process for EVs is usually modeled as an optimization problem such as Integer Linear Programming (ILP), Mixed Integer Linear Programming (MILP), Robust Optimization (RO) and so on. The above optimization problems can be solved with solver software where CPLEX, LINGO, GLPK and IPsolve are the most common tools. However, these tools are not always applicable considering the computational complexity with the increasing of EVs and CSs. In this scenario, the optimization problem is transformed into other forms that is studied based on matching theory, game theory and other theoretical methods.

Emmanouil et al. [14] formulated the problem of managing Electric Vehicle (EV) charging at charging points as a Mixed Integer Quadratic Programming (MIQP) problem to find the optimal charging location and charging times for the EVs and solved by CPLEX. To improve the QoE of EV drivers and the reliability of the power grid, Zeng et al. [25] designed a charging scheduling scheme in driving pattern and a cooperative EV charging and discharging scheduling scheme in parking pattern. In driving pattern, the optimization problem is first formulated as an ILP problem and then modeled as a college admission problem in matching theory concerning with the computational complexity. In the parking pattern, the electricity exchange between charging EVs and discharging EVs is modeled as a many-to-many matching model. Pareto Optimal Matching Algorithm (POMA) is accordingly derived.

6.5 Navigation Service Optimization Based on Battery Swapping Mode

6.5.1 Introduction for Battery Swapping Mode

In recent years, the battery swapping service has emerged as a promising energy supplement scheme in alleviating range anxiety for EV drivers. In this mode, the EV drives on the road and heads into the Battery Swapping Station (BSS) once it requires energy refilling. After the EV arrives the BSS, the depleted battery will be replaced with a full-filled battery immediately.

Different from the plug-in charging model suffers from lengthy charging times, the low residual value of EVs due to battery degradation, stress on the grid during peak periods, etc. The battery swapping service has been developed and regarded as an effective power supplement scheme to mitigate the disadvantage mentioned above. Usually, the arrived EV can be replaced with the available refilled within minutes, which spend similar time with an internal combustion engine (ICE) based vehicles to fill the gasoline tank. Furthermore, a BSS is more favorable than CS for the power system operator because the peak load shifting and relieving stress on the local grid can be achieved by optimizing the charging schedule. Besides, the higher purchasing cost of EVs compared with the ICE-based vehicles hinders the widely popularity. Accordingly, the cost can be reduced by separating the battery from EVs in a battery swapping service. To be specific, consumers buy the EV without batteries and lease batteries from a third-party company. The company has to regularly check and maintain batteries to keep them on the desirable status and increase their lives. To sum up, all participators in the charging service benefit from the battery swapping model.

6.5.2 Challenges for Battery Swapping Service

Although the benefit of the swapping mode has been analyzed, the practical situation is more complex than this. Next, the challenges for battery swapping service is presented from different aspects.

- **Compatibility**: There are many EV manufacturers in the actual operation scenario, and batteries from different brands may not compatible with each other considering size, capability and property. For example, Nissan Leaf uses the proven lithium-manganese (LMO) battery, while other EVs like Tesla uses NCA (nickel, cobalt, aluminum) in the 18,650 cells.
- **Investment**: A huge up-front investment is required on whole battery swapping infrastructure for building a BSS instead of a simple charging pile under plug-in mode. Furthermore, the stock of batteries should exceed by a certain percentage of daily demand to ensure the favorable daily operation of the BSS, which also cost a lot. Therefore, building a nationwide battery swapping system is a great challenge compared with a nationwide EV charging system.
- **Ownership**: In battery swapping service mode, the EV drivers usually do not own the battery and lease it from a third party company. Besides the cost of energy, EV drivers have to pay an additional lease amount and a service charge for each swapping operation. These charges may lead customers for less usage of battery swapping service, especially when a plug-in charging is available.
- **Battery Degradation**: As discussed above, EV drivers do not own a battery and lease it from a third party company. Considering the ownership, EV drivers may not have a good usage habit, which accelerates the battery degradation. Besides,

it is also a huge expenditure for the BSS operator regarding to battery degradation cost.

These challenges widely exist in the development of battery swapping services. However, it also inspires us to find new ideas for navigation service optimization regarding these challenges.

6.5.3 Navigation Service Optimization on Temporal and Spatial Characteristic Under Battery Swapping Mode

Similar to the plug-in charging mode discussed above, the navigation service optimization under the battery swapping mode could also be analyzed from temporal and spatial dimensions. Therefore, we briefly introduce the battery swapping mode with these two aspects in this section.

6.5.3.1 Temporal Dimension

Although the battery swapping model also exhibits similar property with plug-in mode on temporal dimension, we should also notice a huge gap between them is the charging behavior of large-scale stock batteries can be managed and coordinated with various demands. Accordingly, most of these work related to temporal dimension under battery swapping mode is assumed EVs in limited motion and mainly focus on battery swapping management. Sarker et al. [26] considered battery demand uncertainty and proposed an optimization framework for the operating model of BSSs. Infante et al. [27] proposed a two-stage optimization with recourse considering stochastic EV station visits through planning and operations.

6.5.3.2 Spatial Dimension

In spatial dimension, the moving characteristic of EVs is focused and EVs are allocated to the optimal BSSs once requiring battery swapping service. Different from temporal dimension work caring about BSS management and depleted batteries charging scheduling, researches on spatial dimension pay attention to find the optimal station according to the status of BSSs over network. Therefore, the problem of navigation service optimization on spatial dimension mainly focuses on BSS-selection. Ni et al. [28] designed a randomized online algorithm to execute real-time BSS-selection without EV arrival information. Zhang et al. [29] studied the heterogeneity among different types of batteries and proposed a BSS-selection scheme with reservation enable to achieve load balance.

6.5.4 Navigation Service Optimization in Different Manner Under Battery Swapping Mode

6.5.4.1 Centralized Navigation Service Optimization

For the centralized BSS-selection problem, the global information is necessary. Although this approach exhibits high computing costs, it can achieve optimal strategies. All strategies under the centralized manner are made from system-level to obtain global cost function optimization, such as global social welfare. The centralized BSS-selection problem is usually with high computation complexity, nonconvex and nonlinear properties, which is hard to solve through the traditional mathematical analysis method. To address these disadvantages, heuristic optimization algorithms have been regarded as a promising method. A population-based heuristic optimization algorithm was proposed to minimize total charging cost and alleviate power pressure [30]. Besides, the GA has also played a crucial role and made the optimal BSS-selection strategy in a centralized manner by aggregating information from all BSSs. The GA recommended the optimal BSS to EV drivers with the maximum number of available batteries and predicted the minimum waiting time with reservation enabled considering the heterogeneity of batteries [29].

6.5.4.2 Distributed Navigation Service Optimization

For distributed BSS-selection problem, the EV makes battery swapping decisions by individual without global information. Besides, a centralized entity for global scheduling is unnecessary. It is implemented with a lower communication load, but it is hard to reach global optimum compared with centralized manner. Liu et al. [30] have studied the problem the charging and the logistics of depleted batteries and well-charged batteries, which is solved in a distributed way to achieve revenue maximization. Besides, a Mobile Edge Computing (MEC) driven battery swapping service management scheme was also proposed to provide desirable QoE for EV drives from ICT perspective in distributed manner [31].

6.6 New Charging Service Methods for Electric Vehicle

In the above section, we discussed the navigation service optimization for EV from two mainstream methods, including the plug-in charging mode and the battery swapping model. With the development of service optimization for EV, some new charging methods have appeared in the industry at present. In this section, we introduce some new charging service methods for EVs.

- **Mobile Power Unit (MPU)**: The MPU mode has attracted more and more attention in the past decades. As a distributed energy storage system or emergency power supply, the MPU can be regarded as a small "charging station", which can charge the EV anytime and anywhere and share the pressure for the power grid. The MPU is equipped with a large energy storage system and a conventional charging device. This kind of charging mode has a high one-time production cost and can provide fast charging service for multiple electric vehicles at the same time. At present, this model is difficult to be popularized and applied because of its high cost, non-standardization and immature profit model, but it is expected to be mature and commercial in the next few years in terms of policy support, manufacturer's attention and user's requirement.

- **Wireless Charging**: Wireless charging technology for EV mainly realizes the non-contact transmission of power through electromagnetic induction, magnetic resonance and radio wave. Electromagnetic induction wireless charging is currently the most widely used in EV among these methods. Based on electromagnetic induction, a coil is installed at the end of the wireless charging device and the end of the electric vehicle respectively. When AC of a certain frequency is connected to the primary coil, a changing magnetic field will be generated, and the nearby secondary coil will generate a certain induced electromotive force under the action of the changing magnetic field. The conversion efficiency of transferring the electric energy from the wireless charging device to EV is high, but it can only be charged at a single point, and the distance of electric energy transmission is short.

- **V2X**: Because the EV is equipped with a powerful battery, the EV itself can also be regarded as a charging infrastructure. From this point of view, there will still be a certain market for using the EV to discharge the outside world. According to the different objects receiving electricity, V2X technology is derived. V2V means that the receiving objects are also EV, i.e. vehicle to vehicle. Besides, V2X also includes V2L (vehicle to load) and V2H/B (vehicle to home/buildings). V2L (vehicle to load) uses EVs as a mobile power supply to discharge for the third party, such as an electric lamp, electric fan, electric barbecue rack, etc. V2H/B (vehicle to home/buildings) refers to the electric energy interaction between EVs and residential/commercial buildings. In case of power failure, EVs are used as the emergency power supply of homes/public buildings to supply power to important equipment.

References

1. K.L. Hannes Hartenstein, *VANET vehicular applications and inter networking technologies* (Wiley, 2010)
2. ITU-T, Functional requirements and architecture of the NGN release 1 (2006)
3. M.E. Khodayar, L. Wu, M. Shahidehpour, Hourly coordination of electric vehicle operation and volatile wind power generation in SCUC. IEEE Trans. Smart Grid **3**(3), 1271–1279 (2012)
4. Q. Li, T. Cui, R. Negi, F. Franchetti, M.D. Ilic, On-line decentralizedcharging of plug-in electric vehicles in power systems. Mathematics (2011)

5. E. Sortomme, M.M. Hindi, S.D.J. MacPherson, S.S. Venkata, Coordinated charging of plug-in hybrid electric vehicles to minimize distribution system losses. IEEE Trans. Smart Grid **2**(1), 198–205 (2011)
6. C.K. Wen, J.C. Chen, J.H. Teng, P. Ting, Decentralized plug-in electric vehicle charging selection algorithm in power systems. IEEE Trans. on Smart Grid **3**(4), 1779–1789 (2012)
7. Y. He, B. Venkatesh, L. Guan, Optimal scheduling for charging and discharging of electric vehicles. IEEE Trans. Smart Grid **3**(3), 1095–1105 (2012)
8. W. Su, M.Y. Chow, Performance evaluation of an EDA-based largescale plug-in hybrid electric vehicle charging algorithm. IEEE Trans. Smart Grid **3**(1), 308–315 (2012)
9. S. Deilami, A.S. Masoum, P.S. Moses, M.A.S. Masoum, Real-time coordination of plug-in electric vehicle charging in smart grids to minimize power losses and improve voltage profile. IEEE Trans. Smart Grid **2**(3), 456–467 (2011)
10. H.J. Kim, J. Lee, G.L. Park, Constraint-based charging scheduler design for electric vehicles, in *Processing 3rd Asian Conference on Intelligent Information and Database Systems* (Springer, Berlin, Heidelberg, 2012), pp. 266–275
11. R. Jin, B. Wang, P. Zhang, P.B. Luh, Decentralised online chargingscheduling for large populations of electric vehicles: a cyber-physical system approach. Int. J. Parallel Emerg. Distrib. Syst., 29–45 (2013)
12. L. Yao, H. Wei, S. Teng, A real-time charging scheme for demand response in electric vehicle parking station. IEEE Trans. Smart Grid **8**(1), 52–62 (2016)
13. E.S. Rigas, S.D. Ramchurn, N. Bassiliades, G. Koutitas, Congestion management for urban EV charging systems, in *2013 IEEE International Conference on Smart Grid Communications (SmartGridComm)* (IEEE, 2013)
14. S. Wang, S. Bi, Y. Zhang, J. Huang, Electrical vehicle charging station profit maximization: admission, pricing, and online scheduling. IEEE Trans. Sustain. Energy **9**(4), 1722–1731 (2018)
15. J. Mukherjee, A. Gupta, A review of charge scheduling of electric vehicles in smart grid. IEEE Syst. J. **9**(4), 1541–1553 (2015)
16. M. Gharbaoui, L. Valcarenghi, R. Brunoi, B. Martini, M. Conti, P. Castoldi, An advanced smart management system for electric vehicle recharge, in *2012 IEEE International Electric Vehicle Conference* (IEEE, 2012) , pp. 1–8
17. J. Lee, G. Park, Evaluation of a tour-and-charging scheduler for electric vehicle network services, in *International Conference on Computational Science and Its Applications* (Springer, Berlin, Heidelberg, 2013), pp. 110–119
18. Z. Tian, T. Jung, Y. Wang, F. Zhang, L. Tu, C. Xu, X. Li, Real-time charging station recommendation system for electric-vehicle taxis. IEEE Trans. Intell. Transp. Syst. **17**(11), 3098–3109 (2016)
19. Q. Guo, S. Xin, H. Sun, Z. Li, B. Zhang, Rapid-charging navigation of electric vehicles based on real-time power systems and traffic data. IEEE Trans. Smart Grid **5**(4), 1969–1979 (2014)
20. J. Timpner, L. Wolf, Design and evaluation of charging station scheduling strategies for electric vehicles. IEEE Trans. Intell. Transp. Syst. **15**(2), 579–588 (2014)
21. Z. Wei, Y. Li, Y. Zhang, L. Cai, Intelligent parking garage EV charging scheduling considering battery charging characteristic. IEEE Trans. Industr. Electron. **65**(3), 2806–2816 (2018)
22. Y. Cao, T. Wang, O. Kaiwartya, G. Min, N. Ahmad, A. Abdullah, An EV charging management system concerning drivers' trip duration and mobility uncertainty. IEEE Trans. Syst. Man Cybern. Syst. **48**(4), 596–607 (2018)
23. A. Alsabbagh, C. Ma, Distributed charging management of electric vehicles considering different customer behaviors. IEEE Trans. Industr. Inf. **16**(8), 5119–5127 (2020)
24. Y. Yang, Q. Jia, X. Guan, X. Zhang, Z. Qiu, G. Deconinck, Decentralized EV-based charging optimization with building integrated wind energy. IEEE Trans. Autom. Sci. Eng. **16**(3), 1002–1017 (2019)
25. M. Zeng, S. Leng, Y. Zhang, J. He, QoE-aware power management in vehicle-to-grid networks: a matching theoretic approach. IEEE Trans. Smart Grid **9**(4), 2468–2477 (2018)
26. M. Sarker, H. Pandžić, M. Ortega-Vazquez, Optimal operation and services scheduling for an electric vehicle battery swapping station. IEEE Trans. Power Syst. **30**(2), 901–910 (2014)

27. W. Infante, J. Ma, X. Han, A. Liebman, Optimal recourse strategy for battery swapping stations considering electric vehicle uncertainty. IEEE Trans. Intell. Transp. Syst. **21**(4), 1369–1379 (2020)
28. L. Ni, B. Sun, X. Tan, D. Tsang, Inventory planning and real-time routing for network of electric vehicle battery-swapping stations. IEEE Trans. Transp. Electrif. **7**(2), 542–553 (2021)
29. X. Zhang, Y. Cao, L. Peng, N. Ahmad, L. Xu, Towards efficient battery swapping service operation under battery heterogeneity. IEEE Trans. Veh. Technol. **69**(6), 6107–6118 (2020)
30. Q. Kang, J. Wang, M. Zhou, A. Ammari, Centralized charging strategy and scheduling algorithm for electric vehicles under a battery swapping scenario. IEEE Trans. Intell. Transp. Syst. **17**(3), 659–669 (2016)
31. Y. Cao, X. Zhang, B. Zhou, X. Duan, D. Tian, X. Dai, MEC intelligence driven electro-mobility management for battery switch service. IEEE Trans. Intell. Transp. Syst. **22**(7), 4016–4029 (2021)

Chapter 7
AI-Based GEVs Mobility Estimation and Battery Aging Quantification Method

Shuangqi Li and Chenghong Gu

Abstract The bi-directional link between the electrical system and electric vehicles allows vehicle batteries to provide balancing services for the system in a flexible, low-cost, and quick-response manner. However, two critical issues should be solved in realising the benefits of vehicles to grid (V2G) services. Firstly, grid-connected electric vehicle (GEV) mobility may cause uncertainties in the grid's energy storage capacity, which may further impact the power quality and stability of the power grid. Thus, in V2G scheduling, it is necessary to access electric vehicle (EV) mobility and estimate its schedulable capacity and charging requirements (SC&CR) information in advance. Furthermore, the key factor that keeps GEVs owners from becoming the prosumers of the grid is the battery life loss caused by additional operating cycles in V2G service, as well as the concern about expensive battery deterioration costs. To promote the adoption of V2G services, battery life loss should be evaluated and mitigated through a behaviour management algorithm. This chapter investigates and compares the performance of existing V2G capacity prediction methods, including statistical model, learning-based model, and rolling prediction model. Thereafter, it introduces a life loss quantification model to analyse battery aging characteristics when providing V2G services. With the predicted GEVs mobility information and battery aging cost analysis model, V2G resources can be better utilized by producing more efficient strategies.

Keywords Electric vehicle · Vehicle-to-grid · Deep learning · V2G schedulable capacity · Battery aging model

---

S. Li · C. Gu (✉)
Department of Electronic and Electrical Engineering, University of Bath, Bath BA2 7AY, UK
e-mail: c.gu@bath.ac.uk

S. Li
e-mail: sl2908@bath.ac.uk

© The Author(s), under exclusive license to Springer Nature Singapore Pte Ltd. 2023
Y. Cao et al. (eds.), *Automated and Electric Vehicle: Design, Informatics and Sustainability*, Recent Advancements in Connected Autonomous Vehicle Technologies 3, https://doi.org/10.1007/978-981-19-5751-2_7

Abbreviations

V2G	Vehicle to grid
EVs	Electric vehicles
GEV	Grid-connected electric vehicle
SC&CR	Schedulable capacity and charging requirements
SVR	Support vector regression
LSTM	Long short term memory
RNN	Recurrent neural networks
MC	Monte carlo
RMSE	Root-mean-square error
DoD	Depth of discharge
Crate	Charging and discharging rate
NoC	Number of cycles
CTF	Cycles to failure
RCC	Rain-flow cycle counting

7.1 Introduction and Related Works

With growing concerns with environmental pollution, it has become common knowledge that the electrification of the transportation sector is the future trend in addressing energy and environmental problems [1–3]. The adoption of electric vehicles allows more than 85% reductions in carbon emissions in the electricity sector, resulting in a better environmental benefit [4, 5]. GEVs, or more particularly, power batteries, are viewed as invaders to the power grid by default. However, on the other hand, GEV batteries can be employed as mobile energy storage devices to offer flexible energy storage capacities for the power system [6, 7]. Therefore, it comes to the concept of the V2G that effectively integrates the aggregated GEVs into the grid in recent years [8].

As a flexible energy storage resource, EVs are expected to reduce grid load variations, manage renewable energy uncertainties, enhance grid economics, and minimise grid frequency deviations [9, 10]. In recent years, substantial research has studied the better usage of V2G resources to increase grid efficiency and robustness. However, there are still many fundamental problems yet unsolved: (1) It is challenging to accurately estimate V2G schedulable capacity and charging demand, as the irregular travel behaviour of single V2G participants results in low predictability [11]; (ii) it is challenging to quantify and mitigate battery life losses because the battery could undergo a large number of irregular cycles when providing V2G services.

Electric Vehicles can be used as distributed energy resources to provide ancillary services for the power grid. However, to capitalise on this benefit, a key issue to be addressed first is to predict the V2G schedulable capacity to meet different utility demands of power dispatch [11–13]. The V2G schedulable capacity includes schedulable charging capacity and schedulable discharging capacity. In previous works, the statistical forecasting model is one of the most commonly used methods [14, 15]. This method tries to learn the living habits of the residents in different areas from the fixed traffic behavior database. For example, the national household travel survey data is used in ROLINK. J's work [16], and a day-ahead stochastic probability prediction model based on Monte Carlo simulation is built and proved accuracy. However, the statistical model is only applicable for statically calculating V2G capacity but not suitable for online V2G capacity estimation and scheduling. To overcome the above shortcomings of the probability model, data-driven learning methods were studied by some researchers. A linear regression prediction model is established in R.J. Bessa's work [17], the proposed method is verified by numerical analysis and the results show that the established model can simulate the V2G participant behavior effectively. In a further study, the parallel decision tree regression algorithm is used in Ref. [18]; the real-time operation data test results indicate that the maximum prediction error can be limited to 10%. This chapter summarizes the two most commonly used methods: statistical model and learning-based model for V2G schedulable capacity prediction issue.

Battery degradation is the main reason that keeps the EV customer from being the named prosumer of the grid [19–21]. To encourage the enthusiasm of V2G participants, it is necessary to under battery aging characteristics and suppresses the battery life loss in V2G scheduling [22, 23]. The battery aging mechanism and its lifetime prediction have already been well studied in recent years [24]. Cunsong, Wang et al. [25] investigated deep-learning algorithm-driven battery remaining useful life prediction method, and the long short-term memory neural network was employed in their work to learn the long-term dependencies in the lithium-ion battery degraded capacities. The experiment results indicated that the battery capacity estimation error could be limited to 2.5%. An electrochemical mechanism model was established in David A. Howey's work [26] to quantify grid-connected Lithium-ion battery degradation annual cost, which can predict the battery capacity fade with an error of 5%. Mehdi Jafari et al. [27] proposed a method to quantify the grid-connected Lithium-ion battery degradation phenomenon during V2G services based on the General Arrhenius Equation, and the model accuracy was validated under different scenarios and climates. Based on existing research about vehicle battery aging mechanisms, this chapter further analyses battery aging characteristics and proposes a life loss quantification method for guiding battery anti-aging V2G scheduling.

7.2 V2G Schedulable Capacity Modelling and Prediction

Regardless of the type of V2G application, accurate estimation of V2G capacity is of great significance to ensure efficiency and resiliency in power system. Therefore, this part summarizes the most commonly used V2G schedulable capacity prediction methods in the existing literature.

7.2.1 Modelling GEVs Mobility with Statistical Models

Knowing EV mobility and V2G SC&CR information in advance is an essential precondition for both battery protection and optimal grid power balancing [28, 29]. However, V2G SC&CR is highly related to EV behaviors, making it challenging to predict with conventional methods [30]. In previous work, it is usually modelled as statistical forecasting [31] or a learning model [32]. In a statistical forecasting model, the probability distribution of user travel behavior is generated with historical data, and the characteristic parameters are further extracted from the model to predict the EV charge/discharge behavior [33]. The stochastic model proposed in [34] provides a complete solution to simulate and predict the electricity demand resulting from the charging of EVs. The model effectiveness is validated by using real EVs mobility data. To summarize, the V2G participants behavior simulation is carried out by the following 3 steps in statistical methods:

- Firstly, historical GEVs mobility information, including user travel demand and EV status and battery system parameters, are collected and stored in a database for further feature mining.
- Then, based on the collected V2G participants' historical behavior data, a mathematical model is established by capturing statistical features in the dataset. The established model is used to estimate GEVs mobility and V2G capacity information in the short future.
- Giving specific GEV fleet information, the statistical model estimates the travel time, period, distance, and battery state for each V2G participant, which can be used to guide V2G scheduling.

Statistical methods can model V2G behavior patterns accurately, but the accuracy is usually not high enough to guide V2G scheduling. Most of them are designed for optimal siting of charging parking lots, long-term planning of distribution facilities and sizing of grid energy storage systems with GEVs penetration. The estimation and prediction of the prediction model should be improved to facilitate the use of SC&CR information in V2G scheduling.

7.2.2 Modelling GEVs Mobility with Data-Driven Learning Models

Different from statistical models, GEVs discharging and charging power profiles are directly predicted in the learning-based method. A day-ahead learning model is built in [17] based on a linear regression algorithm to capture the time-dependence features in V2G schedulable capacity sequence, and the SC&CR profile is one-off generated by the established learning model. In learning-based methods, instead of a single EV, the prediction is performed on the aggregate EVs. The V2G capacity profile of aggregated EVs is generated by the following 3 steps [35]:

- **STEP 1:** V2G schedulable discharging capacity is firstly quantified based on the charging habit of GEVs. During quantifying the V2G schedulable discharging capacity of a single EV, it is assumed that GEVs are charged just before leaving. To be detailed, GEVs would feed energy to the grid since connection until its SoC reaches the preset lowest value, and the discharge power is the rated discharge power of the battery.
- **STEP 2:** V2G schedulable charging capacity is quantified based on the real charging requirement of GEVs. To satisfy V2G participants charging demand and grid energy efficiency improvement, it is assumed that GEVs would be required to feedback energy to the grid for peak shaving service until its SoC value reaches the lowest value, and the battery would be charged to the preset SoC at departure just right. To be detailed, GEVs would be charged from the lowest SoC value to the preset value with the rated charging power and charging would be finished just before departure.
- **STEP 3:** Aggregate EVs schedulable power profile generation. To reduce the irregular travel behaviour of a single V2G participant, the V2G power profile of all grid-connected EVs are aggregated together.

The procedure of the V2G dispatchable capacity and charging demand profile generation process can be summarized as the pseudo-code shown in Fig. 7.1a, b.

(a)

Algorithm 1: V2G dispatchable capacity quantification model
Input: EV index i, number of vehicles M, grid-connected SoC value SOC_{in}^i, grid-connected time T_{in}^i, preset minimum SoC value SOC_{min}^i, scheduling period T_A to T_B, preset leaving time T_{out}^i, rate discharging power P_{sj}^i.
Output: V2G dispatchable capacity profile P_{total}^i.
1:    **Part one:** V2G dispatchable capacity quantification of an individual EV
3:     **for** EV index i = 1 to M **do**
4:      Generate a power record matrix P_i^t to record the dispatchable power of an individual EV and a SoC value record matrix SOC_i^t.
5:      **for** Time step $t = T_A$ to T_B
6:       **if** $t > T_{in}^i$ and $t < T_{out}^i$ and $SOC_i^t > SOC_{min}^i$ **then**
7:        $P_i^t \leftarrow P_{jd}^i$
8:        $SOC_i^{t+1} \leftarrow SOC_i^t - \frac{\Delta t \times P_i^t \times \eta^i}{C^i} \times 100$
9:      **else**
10:       $P_i^t \leftarrow 0$
11:       $SOC_i^{t+1} \leftarrow SOC_i^t$
12:      **end if**
13:     **end for**
14:    **end for**
15: **Part two:** Aggregate EVs dispatchable capacity calculation
16:    Generate an aggregate EVs power record matrix P_{total}.
17:    **for** Time step $t = T_A$ to T_B
18:     **for** EV index i = 1 to M **do**
19:      $P_{total} \leftarrow P_{total} + P_i^t$
20:     **end for**
21:    **end for**

(b)

Algorithm 2: V2G charging demand quantification model
Input: EV index i, number of vehicles M, grid-connected SoC value SOC_{in}^i, grid-connected time T_{in}^i, preset minimum SoC value SOC_{min}^i, scheduling period T_A to T_B, preset leaving time T_{out}^i, rate discharging power P_{ce}^i.
Output: V2G charging demand profile P_{same}.
1:    **Part one:** V2G charging demand quantification of an individual EV
3:     **for** EV index i = 1 to M **do**
4:      Generate a power record matrix P_i^t to record the dispatchable power of an individual EV and a SoC value record matrix SOC_i^t.
5:      **for** Time step $t = T_B$ to T_A
6:       **if** $t > T_{in}^i$ and $t < T_{out}^i$ and $SOC_i^t > SOC_{min}^i$ **then**
7:        $P_i^t \leftarrow P_{ce}^i$
8:        $SOC_i^{t-1} \leftarrow SOC_i^t - \frac{\Delta t \times P_i^t \times \eta}{C^i} \times 100$
9:      **else**
10:       $P_i^t \leftarrow 0$
11:       $SOC_i^{t-1} \leftarrow SOC_i^t$
12:      **end if**
13:     **end for**
14:    **end for**
15: **Part two:** Aggregate EVs charging demand calculation
16:    Generate an aggregate EVs power record matrix P_{same}.
17:    **for** Time step $t = T_B$ to T_A
18:     **for** EV index i = 1 to M **do**
19:      $P_{same} \leftarrow P_{same} + P_i^t$
20:     **end for**
21:    **end for**

Fig. 7.1 The procedure of the **a** V2G dispatchable capacity and **b** charging demand profile generation process

7.2.3 Learning Algorithms for V2G Capacity Prediction

Based on the generated V2G dispatchable power and charging demand profile, GEVs mobility and V2G capacity can be estimated. Compared to a statistical method, the time-dependence information existing in V2G schedulable capacity sequence can be utilized in data-driven prediction methods. With the volume of the training data set and the number of iterations is increasing, the prediction model accuracy can be further improved. In previous work, random forest [36], logistic regression [37], and support vector regression (SVR) [38] algorithms have been widely used in capturing time-dependence information in time series. Furthermore, recently developed deep learning algorithms such as Convolutional Neural Network, Autoencoder, Deep-Long-Short-Term Memory (LSTM) neural network [39, 40], etc. bring a bright perspective for further improving V2G capacity prediction accuracy.

Based on the existing prediction algorithm, this part further introduces the two most commonly used learning algorithms: Recurrent Neural Networks (RNN) and Deep-LSTM, for V2G capacity prediction issues. RNN is a kind of deep learning algorithm specializing in mining the temporal structure features in time-series data [41], the architecture of RNN is stacking multiple recurrent neurons together into the network architecture. The main feature of RNN is that both feed-forward connection and internal feedback connection exist between the hidden layer units. This feature enables Deep-RNN to capture the temporal feature in a sequence with any length. Therefore, RNN is naturally more capable of forecasting tasks.

The computational graph and its unfolded topological graph are presented in Fig. 7.2 to demonstrate the working process of a Deep-RNN with N layers. The main function of Deep-RNN is to map the input sequence X into the corresponding sequential output O, and fully expose the time-related features in it. It consists of one input layer, several hidden layers, and one output layer, all of which are fully connected. The parameters of deep RNN are updated following the formulas below:

$$a_1^t = b_{in} + W_1^{sc} \cdot h_1^{t-1} + W_{in}^{ic} \cdot x^t \tag{7.1}$$

$$h_i^t = f_{activation}(a_i^t) \tag{7.2}$$

$$a_i^t = b_i + W_i^{sc} \cdot h_i^{t-1} + W_{i-1,i}^{ic} \cdot h_{i-1}^t \tag{7.3}$$

$$o^t = b_{out} + W_n^{sc} \cdot h_n^{t-1} + W_{out}^{ic} \cdot h_n^t \tag{7.4}$$

where: x^t is the system data input at t time step, o^t is the prediction output, h_i^t is the state of ith network layer at time step t, $f_{activation}$ is the activation function, b_i is the bias. W_i^{sc} and $W_{i,i-1}^{ic}$ are the weight of self-connection and inter-connection, respectively. The state of the neuron at time step t depends on three factors: (1) t time step input x^t or sharing state h_{i-1}^t at time t from $(i-1)$ layer, (2) bias b_i, and (3) sharing states h_i^{t-1} at current network layer from last time step t-1.

7 AI-Based GEVs Mobility Estimation and Battery ...

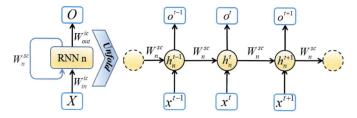

Fig. 7.2 The computational graph and unfolded topological graph of an N layers RNN model

Further, in recent years, LSTM, a specific architecture of RNN, has also been widely used in the industry for solving complex system state prediction problems. In the LSTM unit [42], the gradient can flow for long durations by creating paths and gates in neurons.

To demonstrate how LSTM can memorize long-term patterns, the structure of an LSTM unit is shown in Fig. 7.3. Apart from the original RNN units, a memory parameter vector st is deployed in LSTM cells to keep the memory information, and three gates are also deployed in the LSTM unit to control the information flow: Input Gate, Forget Gate, and Output Gate. In each time step, the memory parameter in LSTM has three operations: (1) discard useless information from memory vector by Forget Gate; (2) add new information it selected from an input vector and previous sharing parameter vector into memory vector st by Input Gate; (3) decide new sharing parameter vector from memory vector by Output Gate. By this operation mechanism, the sharing memory parameters are passed through different time steps to memorize new information and forget out-of-date memories. Therefore, the sharing memory can keep useful information for a long time and result in RNN performance enhancement.

Fig. 7.3 The structure and topological graph of an LSTM unit

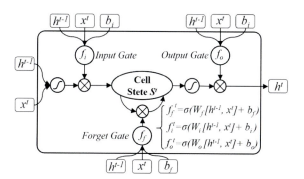

7.2.4 Improving Prediction Accuracy with Rolling Prediction Technology

The rolling prediction technology has been proved to be effective in grid load prediction [43], traffic speed prediction [44, 45], and renewable power generation [46] issues in recent years. According to previous research, the rolling method can significantly improve the accuracy and stability of the established prediction model compared to the offline method [47]. Rolling prediction technology is also effective in improving the V2G capacity prediction model, and the dynamic updated GEVs mobility information can be used as additional input to improve model accuracy. Here, we further provide the flow chart of the rolling V2G schedulable capacity prediction method in Fig. 7.4, which can be divided into four steps [35]:

- Step 1: The travel information data of all V2G participants in a district is obtained and stored in a database, which includes the arriving time T_{in}, departure time T_{out}, arriving battery state SoC_{in}, departing battery SoC requirement SoC_{out}, and the EV types or battery capacity.
- Step 2: The V2G participant behaviour information mentioned in Step 1 is processed based on the V2G schedulable capacity modelling method presented in the previous section, and finally, the aggregate EVs V2G schedulable capacity profiles are formulated.
- Step 3: The V2G schedulable charging and discharging power profile of aggregated EVs generated in Step 2 is used to train the prediction model.

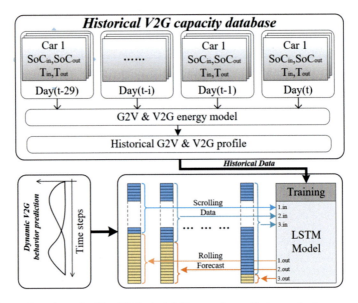

Fig. 7.4 The flow chart of rolling V2G schedulable capacity prediction method

- Step 4: The built prediction model is used to make a rolling prediction on V2G schedulable capacity profile for the next few hours. As shown in Fig. 7.4, historical V2G schedulable charging and discharging capacity data of grid-connected EVs is dynamically updated and used as the rolling input of the prediction model. The prediction process is repeatedly taken out and the model input data is dynamically updated with the latest grid-connected EVs information.

7.2.5 Performance Illustration of Different Methods

This chapter also compares the performance of existing V2G capacity prediction models, including Deep-LSTM, RNN, SVR, statistical Monte-Carlo (MC), and Day-ahead prediction method. The EV travel behavior data of a community with 40 households is adopted in the simulation, and the data from days 1 to 80 and 80 to 85 are used for model training and model verification. The generated V2G schedulable capacity profile of 30 GEVs is shown in Fig. 7.5. The V2G schedulable capacity and charging demand of aggregated GEVs can be successfully modelled, and a regular and predictable profile is generated. With the Deep-LSTM algorithm, the temporal features in the V2G capacity profile can be effectively utilized, and the maximum prediction error of V2G schedulable charging and discharging power can be limited to 5.8% and 4.7% in the simulate period, respectively. Furthermore, with the rolling prediction method, the predictor can almost be kept stable in the whole prediction period, which indicates that the error accumulation phenomenon can be successfully avoided.

The root-mean-square error (RMSE) of the above five prediction methods are illustrated in Fig. 7.6. The prediction error of the statistical MC model-based V2G schedulable discharging power prediction method is 15.71, which indicates that GEVs behavior pattern information can be simulated. The reason is that the static V2G behavior database is used in the MC method, the change of residents' living habits (the weather, the temperature and the traffic condition, etc....) is not

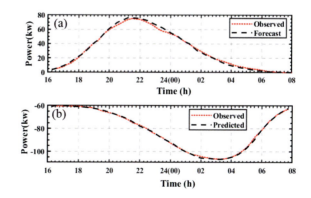

Fig. 7.5 The result of data-driven learning based V2G schedulable **a** discharging power and **b** charging power profile

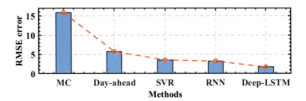

Fig. 7.6 The comparison of prediction accuracy of different algorithms

reflected. Compared to the statistical MC model, the forecasting accuracy of the day-ahead learning prediction model is improved significantly, the prediction RMSE in the whole simulation period is reduced to 5.76, which highlights the effectiveness of the learning-based prediction model. After the rolling prediction technology is adopted, the dynamic hour-ahead time-dependence information in the V2G capacity profile can be adequately utilized. The RMSE of the SVR and RNN algorithm-based prediction method can be successfully reduced to 3.53 and 3.21(38.5% and 46.5% lower than the day-ahead learning model), respectively. The Deep-LSTM algorithm can further excavate the deep time-dependence features in the V2G capacity time series, and the prediction accuracy further obtains a 48.3% improvement.

To summarize, in the existing five V2G capacity prediction methods, the performance of learning-based methods is generally higher than the statistical model. The rolling prediction technology can further improve the accuracy of the prediction model. With rolling prediction technology, both SVR, RNN, and Deep-LSTM models can simulate GEVs charging behaviour accurately. The Deep-LSTM algorithm shows the most outstanding and stable performance in the studied five methods, which provides an effective engineering solution for V2G capacity prediction.

7.3 Quantification of Battery Aging Cost in V2G Management

To quantitively and reasonably mitigate battery aging cost in V2G services, an accurate and specific reference battery life loss label should be provided for the regulator that manages the charging behavior of GEVs. Based on existing work about battery aging mechanisms, this part analyzes and quantifies the vehicle battery life loss in V2G services.

7.3.1 Quantification of Battery Life Loss in V2G Services

According to previous work [48, 49], battery life loss is mainly influenced by the depth of discharge (DoD), the deeper the battery is discharged, the more the life will be consumed. Literature [50] further points out that battery life is also greatly impacted by the charging and discharging rate (Crate). The larger the battery discharging and

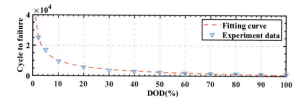

Fig. 7.7 A typical Cycles to Failure profile of the vehicle battery pack [52]

charging current, the faster the battery life will be depleted. Furthermore, [51] also indicates that the number of cycles (NoC) also greatly influences battery life, and both shallow and deep cycles can result in capacity loss. Different batteries have different aging characteristics in different working conditions. The quantification of battery degradation cost is difficult in V2G applications because it may experience several irregular cycles, affecting battery lifespan in different ways and degrees.

Based on the above discussion about the common aging features, including NOC, DOD, and Crate, this section analyses vehicle battery pack degradation characteristics in V2G. The depth of discharge impacts the battery life span mostly; the Cycles to Failure (CTF) provided by the battery manufacturer is one of the most commonly used methods to describe the degradation characteristics of the battery. With the CTF profile, the relationship between maximum life cycles and the DOD of the battery can be derived. This chapter introduces the CTF method to transform the extracted battery cycle information to the corresponding capacity degradation degree. A typical CTF profile of the battery in GEVs is shown in Fig. 7.7.

The curve fitting method is commonly used in existing work to make the CTF profile continuous, where Gaussian function is used to approximate the relationship between the CTF and the battery DOD:

$$CTF = f_{CTF}(DOD) = a_1 e^{-\frac{DOD-b_1}{c_1}^2} + a_2 e^{-\frac{DOD-b_2}{c_2}^2} \quad (7.5)$$

where: a_1 to c_2 are the curve fitting coefficients in the studied typical CTF profile, a1 = 4.254 × 10^{43}, b1 = −10.16, c1 = 1.07, a2 = 2.134 × 10^{29}, b2 = −63.13, c2 = 8.235.

According to previous literature about battery characteristics, the discharging current also impacts its life greatly. Therefore, the Crate is usually considered in most battery aging models as well. The most commonly used method to quantify the influence of Crate on battery life is using a nonlinear correction index to amend the CTF profile. According to the previous research about the battery aging model, the relationship between the capacity retention and the Crate can be described as:

$$\nabla(\Delta Capacity) = d_0 + d_1 \cdot e^{-\frac{Crate-d_2}{d_3}^2} \quad (7.6)$$

where d_0 to d_3 are the curve fitting coefficients, the values of which are $d_0 = 0.880$, $d_1 = 0.093$, $d_2 = -0.064$, $d_3 = -1.378$. The larger the current battery is discharged, the more life will be consumed. The capacity retention of the battery is also proportional to the NoC and DoD, which can be described by the following equation:

$$\nabla(\Delta Capacity) = f_{CA}(CTF) = f_0 + f_1 \cdot CTF \tag{7.7}$$

where f_0 to f_1 are the curve fitting coefficients, according to the previous literature, the values of them are $f_0 = -0.0018, f_1 = 0.96$.

As a result, the equivalent battery cycles can be derived by combining the Eqs. (7.2)–(7.4):

$$CTF_E = f_{CTF}(DOD) \cdot \frac{1}{f_{CA}(\nabla(\Delta Capacity))} \tag{7.8}$$

Different from conventional electric vehicle operation conditions, the battery may experience several cycles in V2G service, to enable the additivity of the quantified battery equivalent battery cycles, the equivalent battery aging factor η, which reflects the relationship between the battery experienced cycles and life state can be defined as:

$$\eta(k) = \frac{1}{CTF_E(k)} \tag{7.9}$$

where: k means the index of the extracted battery cycles. According to the total number of cycles and DOD of the battery, the total life loss in V2G service can be calculated as:

$$LL = \sum_{k=1}^{k=n} \eta(k) \tag{7.10}$$

The value of LL is between 0 and 1, and the higher the value of LL, the more the battery life will be exhausted in V2G. Here, the value of LL is used as the aging index for guiding V2G behaviour management.

The quantified battery life loss under different DODs and Crates in CTF based battery aging model is shown in Fig. 7.8 [53]. With the increasing of battery DoD and Crate, battery equivalent battery life cycle generally reduced, which indicated that battery life loss increases with DoD and Crate. The shallow cycles with low Crate show very limited influence on battery life, while the deep cycles with high Crate greatly impact battery life. Based on the above analysis, deep cycles and high Crate working conditions should be avoided to protect the vehicle battery; meanwhile, the vehicle battery sum of NoC should also be limited to reduce accumulated life loss.

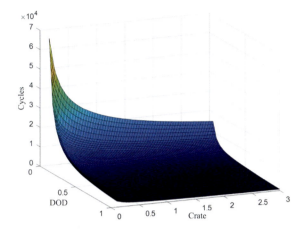

Fig. 7.8 The quantified battery life loss responding profile under different DODs and Crates

7.3.2 Quantification of Battery Aging Cost in V2G Services

Based on the introduced battery aging model previously, battery life loss can be quantitively analyzed according to its NOC, DOD, and Crate information. This part further introduces a battery aging cycles extraction method to analyze aging cycles information in V2G profiles. The rain-flow cycle counting (RCC) algorithm has been identified as one of the most effective methods in the mechanical fault diagnosis and identification issue [54, 55]. This chapter introduces the RCC algorithm applied in battery aging trajectory extraction to extract irregular charging and discharging cycles that the battery experiences caused by V2G. Basically, the counting of battery life cycles can be achieved by using the RCC algorithm through the following three steps:

- Firstly, battery discharging/charging trajectory data is pre-processed by searching for adjacent data points with reverse polarity so that the local SoC maxima and minima can be found and stored in a matrix.
- Secondly, the discharging and charging sub-cycles are extracted by analysing the turning points of the processed SoC profile, and the amplitudes of these sub-cycles are summed together to obtain the corresponding full cycles.
- At last, the battery number of cycles and the corresponding DOD data are extracted and stored for further analysis.

To facilitate the reproduction of the above process, we provide the pseudo-code of the RCC algorithm-based V2G aging cycles extraction method, shown in Fig. 7.9. The battery degradation can be quantified by further analyzing the extracted NOC, DOD, and Crate information from different cycles in the V2G power profile.

Algorithm: Rain-flow counting algorithm-based battery aging trajectory extraction method	
Input: Battery SOC trajectory in the V2G	
Output: Battery cycles, corresponding DODs	

```
 1: Part one: SoC data reconstructing
 2:     Store SOC data in A matrix and B matrix, and the length of the SOC data is donated as m
 3:         for i = 2 to m-1 do
 4:             if (A (i-1) - A (i)) * (A (i) - A (i+1)) > 0 then
 5:                 A (i) ← Ø
 6:                 B ← A
 7:             end if
 8:         end for
```

```
 9: Part two: Definition the function of judging full cycle flag(X)
10:     n ← length of matrix B
11:         for j = 1 to n-4 do
12:                 s1 ← | B (j+1) – B (j+2) |
13:                 s2 ← | B (j+3) – B (j) |
14:                     if s1 <= s2 do
15:                         flag ← 1, break
16:                     else
17:                         flag ← 0, continue
18:                     end if
19:         end for
20:     return flag
```

```
21: Part three: Calculate the amplitude of each full cycle
22:     Store the cycle amplitude matrix F.
23:         while flag (B)==1 or flag(B)==0 do
24:             if flag (B) ==1 do
25:                 for j ← 1 to n-4 do
26:                         s1 ← | B (j+1) – B (j+2) |
27:                         s2 ← | B (j+3) – B (j) |
28:                             if s1 < s2 do
29:                                 F ← S1
30:                                 B(j+1) ←Ø
31:                             else
32:                                 Continue
33:                             end if
34:                 end for
35:             else if flag (B) == 0
36:                 Continue
37:             end if
38:         end while
```

Fig. 7.9 RCC algorithm-based V2G aging cycles extraction method

7.4 Conclusion

This chapter performs a thorough literature review of state-of-the-art research works in terms of GEVs mobility estimation, prediction, and battery aging cost evaluation in V2G services. By introducing the statistical, data-driven, and rolling prediction methods, GEVs mobility and V2G capacity can be quantitatively estimated and predicted. The predictive results provide detailed prior information for guiding V2G behaviour management and can further promote the reasonable utilization of GEVs energy storage capacity. To illustrate the effectiveness of V2G capacity prediction, the charging behaviour of 40 GEVs within 90 days is used to simulate V2G system operation and generate V2G power profiles. Case study results validate the predictability of V2G schedulable capacity and charging demand profile. Afterwards, based on the battery aging characteristic analysis results in previous literature, the battery life loss phenomenon in V2G services is quantitatively analysed. The introduced battery aging model can be used to guide anti-aging GEVs charging scheduling by providing a life loss feedback signal. With the predicted GEVs mobility information and battery aging cost analysis model, V2G resources can be better utilized by producing more efficient strategies.

References

1. E. Taherzadeh, M. Dabbaghjamanesh, M. Gitizadeh, A. Rahideh, A new efficient fuel optimization in blended charge depletion/charge sustenance control strategy for plug-in hybrid electric vehicles. IEEE Trans. Intell. Veh. 3(3), 374–383 (2018)
2. A. Maroosi, A. Ahmadi, A.E. Nezhad, H.A. Shayanfar, Comment on "Resource Scheduling Under Uncertainty in a Smart Grid With Renewables and Plug-In Vehicles" by A. Y. Saber and G. K. Venayagamoorthy. IEEE Syst. J. 10(1), 147–150 (2016)
3. S. Li, C. Gu, P. Zhao, S. Cheng, A novel hybrid propulsion system configuration and power distribution strategy for light electric aircraft. Energy Convers. Manag. 238, 114171 (2021)
4. N. Mizuta, Y. Susuki, Y. Ota, A. Ishigame, Synthesis of spatial charging/discharging patterns of in-vehicle batteries for provision of ancillary service and mitigation of voltage impact. IEEE Syst. J. 13(3), 3443–3453 (2019)
5. S. Li, H. He, P. Zhao, Energy management for hybrid energy storage system in electric vehicle: a cyber-physical system perspective. Energy 230, 120890 (2021)
6. L. Jin, C.-K. Zhang, Y. He, L. Jiang, M. Wu, Delay-dependent stability analysis of multi-area load frequency control with enhanced accuracy and computation efficiency. IEEE Trans. Power Syst. (2019)
7. Q. Yang et al., An improved vehicle to the grid method with battery longevity management in a microgrid application. Energy 198, 117374 (2020)
8. J.C. Mukherjee, A. Gupta, A review of charge scheduling of electric vehicles in smart grid. IEEE Syst. J. 9(4), 1541–1553 (2015)
9. H. Fan, C. Duan, C.-K. Zhang, L. Jiang, C. Mao, D. Wang, ADMM-based multiperiod optimal power flow considering plug-in electric vehicles charging. IEEE Trans. Power Syst. 33(4), 3886–3897 (2018)
10. A. Dutta, S. Debbarma, Frequency regulation in deregulated market using vehicle-to-grid services in residential distribution network. IEEE Syst. J. 12(3), 2812–2820 (2018)
11. A. Azizivahed, E. Naderi, H. Narimani, M. Fathi, M.R. Narimani, A new bi-objective approach to energy management in distribution networks with energy storage systems. IEEE Trans. Sustain. Energy 9(1), 56–64 (2018)
12. F. Ye, Y. Qian, R.Q. Hu, Incentive load scheduling schemes for PHEV battery exchange stations in smart grid. IEEE Syst. J. 11(2), 922–930 (2017)
13. Y. Cao, N. Wang, G. Kamel, Y. Kim, An electric vehicle charging management scheme based on publish/subscribe communication framework. IEEE Syst. J. 11(3), 1822–1835 (2017)
14. L. Agarwal, W. Peng, L. Goel, Probabilistic estimation of aggregated power capacity of EVs for vehicle-to-grid application, in *2014 International Conference on Probabilistic Methods Applied to Power Systems (PMAPS)* (IEEE, 2014), pp. 1–6

15. Y. Sun, H. Yue, J. Zhang, C. Booth, Minimisation of residential energy cost considering energy storage system and EV with driving usage probabilities. IEEE Trans. Sustain. Energy, 1 (2018)
16. J. Rolink, C. Rehtanz, Large-scale modeling of grid-connected electric vehicles. IEEE Trans. Power Delivery **28**(2), 894–902 (2013)
17. R.J. Bessa, M. Matos, Global against divided optimization for the participation of an EV aggregator in the day-ahead electricity market. Part I: theory. Electr. Power Syst. Res. **95**, 309–318 (2013)
18. M. Mao, Y. Wang, Y. Yue, L. Chang, Multi-time scale forecast for schedulable capacity of EVs based on big data and machine learning, in *2017 IEEE Energy Conversion Congress and Exposition (ECCE)* (IEEE, 2017), pp. 1425–1431
19. O. Kolawole, I. Al-Anbagi, Electric vehicles battery wear cost optimization for frequency regulation support. IEEE Access **7**, 130388–130398 (2019)
20. T. Mao, B. Zhou, X. Zhang, Accommodating discharging power with consideration of both EVs and ESs as commodity based on a two-level GA algorithm. IEEE Access (2019)
21. C. Guenther, B. Schott, W. Hennings, P. Waldowski, M.A. Danzer, Model-based investigation of electric vehicle battery aging by means of vehicle-to-grid scenario simulations. J. Power Sour. **239**, 604–610 (2013)
22. C. Li et al., An optimal coordinated method for EVs participating in frequency regulation under different power system operation states. Ieee Access **6**, 62756–62765 (2018)
23. A. Ahmadian, M. Sedghi, B. Mohammadi-ivatloo, A. Elkamel, M.A. Golkar, M. Fowler, Cost-benefit analysis of V2G implementation in distribution networks considering PEVs battery degradation. IEEE Trans. Sustain. Energy **9**(2), 961–970 (2017)
24. S. Li, H. He, J. Li, Big data driven lithium-ion battery modeling method based on SDAE-ELM algorithm and data pre-processing technology. Appl. Energy **242**, 1259–1273 (2019)
25. C. Wang, N. Lu, S. Wang, Y. Cheng, B. Jiang, Dynamic long short-term memory neural-network-based indirect remaining-useful-life prognosis for satellite lithium-ion battery. Appl. Sci. **8**(11), 2078 (2018)
26. Y. Zhang, X. Du, M. Salman, E. Systems, Battery state estimation with a self-evolving electrochemical ageing model. Int. J. Electr. Power **85**, 178–189 (2017)
27. M. Jafari, A. Gauchia, S. Zhao, K. Zhang, L. Gauchia, Electric vehicle battery cycle aging evaluation in real-world daily driving and vehicle-to-grid services. IEEE Trans. Transp. Electrif. **4**(1), 122–134 (2018)
28. A. Kavousi-Fard, A. Abunasri, A. Zare, R. Hoseinzadeh, Impact of plug-in hybrid electric vehicles charging demand on the optimal energy management of renewable micro-grids. Energy **78**, 904–915 (2014)
29. Y. Wu, J. Zhang, A. Ravey, D. Chrenko, A. Miraoui, Real-time energy management of photovoltaic-assisted electric vehicle charging station by markov decision process. J. Power Sour. **476**, 228504 (2020).
30. J. Zhao, F. Wen, Z.Y. Dong, Y. Xue, K.P. Wong, Optimal dispatch of electric vehicles and wind power using enhanced particle swarm optimization. IEEE Trans. Industr. Inf. **8**(4), 889–899 (2012)
31. T. Zhang, X. Chen, Z. Yu, X. Zhu, D. Shi, A monte carlo simulation approach to evaluate service capacities of EV charging and battery swapping stations. IEEE Trans. Industr. Inf. **14**(9), 3914–3923 (2018)
32. F. Varshosaz, M. Moazzami, B. Fani, P. Siano, Day-ahead capacity estimation and power management of a charging station based on queuing theory. IEEE Trans. Industr. Inf. **15**(10), 5561–5574 (2019)
33. Z. Hu et al., Multi-objective energy management optimization and parameter sizing for proton exchange membrane hybrid fuel cell vehicles. Energy Convers. Manag. **129**, 108–121 (2016)
34. M. Alizadeh, A. Scaglione, J. Davies, K.S. Kurani, A scalable stochastic model for the electricity demand of electric and plug-in hybrid vehicles. IEEE Trans. Smart Grid **5**(2), 848–860 (2014)
35. S. Li, C. Gu, J. Li, H. Wang, Q. Yang, Boosting grid efficiency and resiliency by releasing V2G potentiality through a novel rolling prediction-decision framework and deep-LSTM algorithm. IEEE Syst. J. **15**(2), 2562–2570 (2021)

36. Y. Liu and H. Wu, Prediction of road traffic congestion based on random forest, in *2017 10th International Symposium on Computational Intelligence and Design (ISCID)*, vol. 2 (IEEE, 2017), pp. 361–364
37. M.E. Shipe, S.A. Deppen, F. Farjah, E.L. Grogan, Developing prediction models for clinical use using logistic regression: an overview. J. Thorac. Dis. **11**(Suppl 4), S574 (2019)
38. H. Yang, K. Huang, I. King, M.R. Lyu, Localized support vector regression for time series prediction. Neurocomputing **72**(10–12), 2659–2669 (2009)
39. Q. Zhu, et al., Learning temporal and spatial correlations jointly: a unified framework for wind speed prediction. IEEE Trans. Sustain. Energy, 1 (2019)
40. M. Khodayar, S. Mohammadi, M.E. Khodayar, J. Wang, G. Liu, Convolutional graph autoencoder: a generative deep neural network for probabilistic spatio-temporal solar irradiance forecasting. IEEE Trans. Sustain. Energy, 1 (2019)
41. H. Shi, M. Xu, R. Li, Deep learning for household load forecasting—a novel pooling deep RNN. IEEE Trans. Smart Grid **9**(5), 5271–5280 (2018)
42. T.-Y. Kim, S.-B. Cho, Predicting the household power consumption using CNN-LSTM hybrid networks, in *International Conference on Intelligent Data Engineering and Automated Learning* (Springer, 2018), pp. 481–490
43. S. Wang, X. Wang, S. Wang, D. Wang, Bi-directional long short-term memory method based on attention mechanism and rolling update for short-term load forecasting. Int. J. Electr. Power Energy Syst. **109**, 470–479 (2019)
44. O. Elbagalati, M.A. Elseifi, K. Gaspard, Z. Zhang, Prediction of in-service pavement structural capacity based on traffic-speed deflection measurements. J. Transp. Eng. **142**(11), 04016058 (2016)
45. X. Chen, S. Zhang, L. Li, L. Li, Adaptive rolling smoothing with heterogeneous data for traffic state estimation and prediction. IEEE Trans. Intell. Transp. Syst. **20**(4), 1247–1258 (2018)
46. R. Palma-Behnke, et al., A microgrid energy management system based on the rolling horizon strategy. IEEE Trans. Smart Grid **4**(2), 996–1006 (2013)
47. Z. Cao, Y. Han, J. Wang, Q. Zhao, Two-stage energy generation schedule market rolling optimisation of highly wind power penetrated microgrids. Int. J. Electr. Power Energy Syst. **112**, 12–27 (2019)
48. I. Baghdadi, O. Briat, J.Y. Delétage, P. Gyan, J.M. Vinassa, Lithium battery aging model based on Dakin's degradation approach. J. Power Sour. **325**, 273–285 (2016)
49. S. Li, P. Zhao, Big data driven vehicle battery management method: a novel cyber-physical system perspective. J. Energy Storage **33**, 102064 (2021)
50. J. Schmalstieg, S. Käbitz, M. Ecker, D.U. Sauer, A holistic aging model for Li (NiMnCo) O2 based 18650 lithium-ion batteries. J. Power Sour. **257**, 325–334 (2014)
51. M. Petit, E. Prada, and V. Sauvant-Moynot, Development of an empirical aging model for Li-ion batteries and application to assess the impact of Vehicle-to-Grid strategies on battery lifetime. Appl. Energy **172**, 398–407 (2016)
52. Q. Badey, G. Cherouvrier, Y. Reynier, J.M. Duffault, S.J.C.T.E. Franger, Ageing forecast of lithium-ion batteries for electric and hybrid vehicles. Curr. Top. Electrochem **16**, 65–79 (2011)
53. S. Li, H. He, C. Su, P. Zhao, Data driven battery modeling and management method with aging phenomenon considered. Appl. Energy **275**, 115340 (2020)
54. S. Li, J. Li, C. Su, Q. Yang, Optimization of bi-directional V2G behavior with active battery anti-aging scheduling. IEEE Access **8**, 11186–11196 (2020)
55. M. Köhler, S. Jenne, K. Pötter, H. Zenner, Comparison of the counting methods for exemplary load time functions, in *Load Assumption for Fatigue Design of Structures and Components* (Springer, 2017), pp. 85–91.

Chapter 8
Multi-objective Bi-directional V2G Behavior Optimization and Strategy Deployment

Shuangqi Li and Chenghong Gu

Abstract The mitigation of peak-valley difference and power fluctuations are of great significance to the economy and stability of the power grid. The concept of the vehicle to grid (V2G) technology makes it possible to integrate electric vehicles (EVs) into the grid as distributed energy resources and provide power balancing service to the grid. This chapter introduces a V2G scheduling approach that can provide power balancing services to the grid while mitigating the battery aging phenomenon. Firstly, an intelligent V2G behaviour management framework is presented, which enables the comprehensive utilization of prediction information in V2G scheduling. Then, a commonly used multi-objective V2G behavior optimization model is introduced, in which minimal battery degradation and grid load fluctuation are the optimization objectives. Meanwhile, a multi-population collaborative mechanism, which is particularly designed for the V2G scheduling problem and has been proved effective in previous literature, is also introduced to improve the performance of the heuristic optimization-based V2G scheduling model. Two commonly used real-time strategy deployment methods: fuzzy logic and neural network, are further introduced for online V2G scheduling. With the presented methods, grid-connected electric vehicle (GEV) energy storage capacity can be scheduled to provide power balancing services to the grid while significantly mitigating battery aging.

Keywords Electric vehicle · Vehicle-to-grid · Optimization · Power balancing · Battery aging mitigation

---

S. Li · C. Gu (✉)
Department of Electronic and Electrical Engineering, University of Bath, Bath BA2 7AY, UK
e-mail: c.gu@bath.ac.uk

S. Li
e-mail: sl2908@bath.ac.uk

© The Author(s), under exclusive license to Springer Nature Singapore Pte Ltd. 2023
Y. Cao et al. (eds.), *Automated and Electric Vehicle: Design, Informatics and Sustainability*, Recent Advancements in Connected Autonomous Vehicle Technologies 3, https://doi.org/10.1007/978-981-19-5751-2_8

Abbreviations

V2G	Vehicle to grid
EVs	Electric vehicles
GEV	Grid-connected electric vehicle
PSO	Particle swarm optimization
DOD	Depth of discharge
ICT	Information and communication technology
FA-CD	Future arrival EV's charging and discharging capacity
MC-PSO	Multi-population collaborative PSO
STD	Standard deviation

8.1 Introduction

Electric vehicles and power grids are the two important components of the future low-carbon energy system. Instead of a one-way energy flow from the grid to EVs, its bi-directional link allows the vehicle batteries to be used in the power grid in a flexible, cost-effective, and quick-response manner [1, 2]. As a result, the notion of V2G emerges, which integrates EVs as distributed energy resources into the grid [3–5]. As a kind of variable capacity energy storage device, EVs and their battery system is primarily used for reducing grid load variations [6], managing renewable energy sources [7], and minimising grid frequency fluctuations [8]. Mukesh Singh et al. [9] proposed a V2G scheduling method for grid voltage support and peak demand management that utilises a fuzzy logic controller. Results demonstrated that power levelling and peak shaving targets could be accomplished by deferring EV charging to off-peak hours and discharging energy back to the grid during peak hours. Kristien Clement-Nyns et al., [10] propose an online GEV charging coordination technique based on a dynamic programming algorithm that optimises charging profile of GEVs by reducing power loss.

Currently, the existing V2G scheduling method can be divided into day-ahead offline mode and real-time online mode two categories. In the day-ahead method, the charging behavior of GEVs is offline scheduled based on the estimated grid and GEVs states information [11–13]. Ahmet DOGAN et al. [14] came up with a coordination scheme based on a Genetic Algorithm, in which the status of each EV is decided to minimize the coordination cost considering network and EV constraints. Real-time online optimization is different from day-ahead offline optimization in two ways: first, the information needed in optimization is the real-time status of grid-connected EVs, rather than the prediction result; second, the optimization result is the current charge/discharge power but not the control sequence for next few hours [15, 16]. Wu et al. [17] proposed a dynamic online approach for vehicle charging coordination based on Particle Swarm Optimization (PSO) algorithm. The V2G behavior of each grid-connected EV in the future 5 min is scheduled in real-time, which successfully shifts charging demands to off-peak hours and reduces the overall cost.

The concern of the battery degradation is the main reason that keeps the EV customer from being the named prosumer of the gird [18–20]. To encourage user participation, user profit must be guaranteed by regulating battery degradation in V2G application [21, 22]. Many studies have been conducted to quantify GEV battery aging cost and its impact on the V2G services in recent years. Landi et al. [23] proposed a fuzzy logic-based battery health state assessing method in V2G application, in which the influence of depth of discharge (DOD), temperature, charge/discharge rate on battery degradation are taken into consideration. In [24], V2G scheduling is modelled as a stochastic optimization problem and the mitigation of battery aging is realized by setting the number of cycle constraints. Simulation results on a smart grid system indicated that the total economy of the integrated transportation-energy system could be significantly improved.

This chapter mainly focuses on summarizing the latest methods for solving the following issues in V2G behavior management: (1) The utilization of predictive information, which can help scheduled systems obtain global optimal solutions in the whole temporal domain; (2) multi-objective scheduling algorithm that can provide both power balancing services and mitigate vehicle battery aging; (3) advanced optimization algorithm for solving the large-scale multi-objective optimization problem in V2G scheduling; (4) real-time deployment of optimization-based strategy in real-time V2G scheduling.

8.2 Intelligent V2G Scheduling Systems

In most previous work, V2G management is realized based on Information and Communication Technologies [25, 26] (ICTs), which facilitate the flexible utilization of prediction information and optimization algorithms. As shown in Fig. 8.1, existing intelligent V2G scheduling systems commonly consist of 4 parts: the prediction information module, user information collection, V2G management, and EV Smart charger hardware [27]:

- User information collection module is used to collect the household electricity load and EV's charging demand information by ICT technology, and the information includes EVs accessing time, depart time, arriving SoC, and the expected departure SoC and so on.
- In the information prediction module, historical V2G information Databases, as well as a historical load demand database corresponding to the users in different districts, are established. Meanwhile, future arrival EV's charging demand and discharging capacity (FA-CD) information and future load demand information are predicted based on the aforementioned database and provided as important data foundations for V2G scheduling.

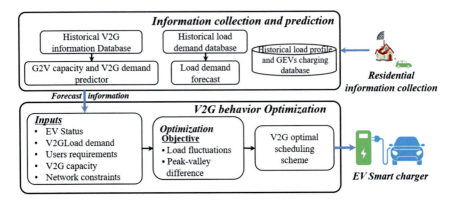

Fig. 8.1 The architecture of intelligent V2G management system

- V2G management module is used to formulate the V2G charge/discharge schemes for every grid-connected EV with both the information of grid-connected EVs and the prediction result. The derived optimal control sequence is sent to EV smart chargers.
- At last, according to the charge/discharge control sequence, EV smart charger controls the charge/discharge power of every EV in real-time by power electronic devices.

8.3 Multi-objective V2G Behaviour Management

In the existing literature, V2G management is commonly modelled as a mathematical optimization problem to derive the optimal strategy [28–30]. However, the inherent high-dimensional, large-scale characteristics of V2G scheduling cannot be neglected [31, 32]. Moreover, in the V2G scheme with the consideration of battery active anti-aging, the objective to minimize grid load fluctuation somehow conflicts with that to minimize battery degradation. The trade-off actually is a problem to search for a Pareto-optimal point and is normally difficult to solve. In addition, the objective function is usually not simply linear or quadratic, so the regular convex optimization method is not suitable in this case [33]. The introduction of quantitative indexes of battery degradation makes it much worse than the objective function is not continuous, non-drivable and non-gradient, where common gradient descent algorithms are not applicable [34].

The PSO algorithm is a typical heuristic algorithm proposed by Kennedy and Eberhart in 1995. At present, the PSO algorithm is widely used in Path Planning [35], optimization design [36] and systems identification [37], etc. The PSO algorithm has already been widely used in V2G management. An improved PSO algorithm is proposed in [38] to optimize grid-connected EVs' charging and discharging behaviors. The simulation results validated the effectiveness and efficiency of the PSO algorithm when dealing with V2G behavior management issues. The PSO algorithm has also been used in the V2G management research conducted by Somayeh [39] and João [40]. This chapter summarises the existing optimization-based V2G scheduling method and introduces a commonly used mathematical optimization model for V2G scheduling issues, in which the minimal battery degradation, grid load, voltage, and frequency fluctuations are designed as the optimization objectives.

The optimization variable in V2G scheduling is commonly selected as the charge/discharge power of every grid-connected EV. The particle dimension is $(n + 1) \times (T_u + T_w)$. Where n is the total number of EVs already in the grid. 1 represents the future available V2G capacity of EVs that will connect to the grid in later control steps. T_w and T_u are the number of decision points in the future and past control step, respectively. The position of the particle is designed as follows:

$$\mathbf{P}_1 = \begin{bmatrix} P_{1,1} & \cdots & P_{1,j} & \cdots & P_{1,n} & P_{1,u+1} & \cdots & P_{1,u+w} \\ P_{2,1} & & P_{2,j} & \cdots & P_{2,u} & P_{2,t+1} & \cdots & P_{2,u+w} \\ \vdots & \ddots & \vdots & \ddots & \vdots & \vdots & \ddots & \vdots \\ P_{i,1} & \cdots & P_{i,j} & \cdots & P_{3,u} & P_{3,u+1} & \cdots & P_{i,u+w} \\ \vdots & \ddots & \vdots & \ddots & \vdots & \vdots & \ddots & \vdots \\ P_{n,1} & \cdots & P_{n,j} & \cdots & P_{n,n} & P_{n,n+1} & \cdots & P_{n,u+w} \\ P_{n+1,1} & \cdots & P_{n+1,j} & \cdots & P_{n+1,u} & P_{n+1,n+1} & \cdots & P_{n+1,n+w} \end{bmatrix} \tag{8.1}$$

where: $P_{i,j}$ represents the power state of EV_i in control step j, $P_{n+1,j}$ reflects the utilization degree of future V2G schedulable capacity. Here, the historical V2G behaviors are stored in the particle; it is not schedulable but directly influences future V2G scheduling.

To estimate EV SoC accurately, the recurrence formula is as follows [41]:

$$SoC_{t+1}^i = SoC_t^i + \frac{\Delta t \times P_t^i \times \eta^i}{C^i} \times 100 \tag{8.2}$$

where Δt is the control time-step, C^i is the battery capacity of EV_i. η^i is the battery charge/discharge efficiency.

The most basic optimization objective in V2G scheduling is to provide load-shifting service, which can be described as to minimize load fluctuation variance [39]:

$$\min, \Delta P = \min \left\{ \frac{1}{u+w} \sum_{t=1}^{u+w} \left[\mathbf{P}_{\mathbf{load}}(\mathbf{t}) + \sum_{i=1}^{n} \mathbf{P}_{\mathbf{I}}(\mathbf{t}) - \overline{P}_{AV} \right]^2 \right\} \tag{8.3}$$

where: $\mathbf{P}_{\mathbf{load}}(\mathbf{t})$ is the system load in the time slot t, $\overline{P}_{AV}$ is the average grid load level.

V2G energy storage capacity can also be used to provide voltage regulation services [42, 43], which can be represented by the following equation to mitigate the voltage fluctuation of the power grid nodes:

$$\min, \Delta V = \min\left\{\left\{\sum_b^{i=1} V_{\text{bus}}^{i,t} - \sum_b^{i=1} V_{\text{bus}}^{i,nom}\right\}\right\} \tag{8.4}$$

where: $V_{\text{bus}}^{i,t}$ and $V_{\text{bus}}^{i,nom}$ are the sampled grid real voltage state and normal node voltage level. Similar to voltage regulation, V2G energy storage capacity can also be used to provide frequency response services to the power grid. In [44], it is modelled as a time-windowed optimal control problem:

$$\dot{x}_G = A_G x_G + B_G u_G + B_{d_G} d_G \tag{8.5}$$

where the $\dot{x}_G = \left[\Delta \dot{f}, \dot{P}_{LB}, \sum \dot{P}_{EVi}, \dot{P}_{\text{load}}, P_{re}\right]^T$, AG is 14×14 matrix first line as $\left[A_G, B_G, \sum_{10} B_G, -B_G, B_G\right]$ and others lines all zero, uG as $\left[\dot{P}_{LB}, \sum \dot{P}_{EVi}, \dot{P}_{\text{load}}, \dot{P}_{re}\right]^T$, dG as the power disturbance inside each energy sectors because of the loss, and simplified by the direct proportion factor ß applied on the exchange power, and BdG is 13×13 matrix first line as $\left[B_G, \sum_{10} B_G, -B_G, B_G\right]$ with others lines all zero.

Apart from grid stability and economy, battery degradation phenomenon resulting by participating in V2G is also considered in the existing literature to improve its total economy:

$$\min, \Delta L = \min\left\{\sum_{i=1}^n N_i^{cycle} + N_i^{h-cycle}\right\} \tag{8.6}$$

N_i^{cycle} and $N_i^{h-cycle}$ are the battery number of cycles and half-cycles of EV_i in V2G scheduling, which can be calculated by the battery life loss presented in Chap. 7.

When formulating V2G strategies, the travel demand of V2G participants should be satisfied, and the battery charging process should be completed before departure [39]:

$$SoC_i^{end} \geq SoC_i^{set} \tag{8.7}$$

Battery life is mainly influenced by the number of cycles, DOD and charge/discharge rate. The number of cycles has been considered in the objective function, the DOD and charge/discharge rate is restricted by the following constraints:

$$-P_{i,disch\,arg}^{\max} \leq P_{i,t} \leq P_{i,ch\,arg\,e}^{\max} \tag{8.8}$$

$$SoC_{\min} \leq SoC_{i,t} \leq SoC_{\max} \tag{8.9}$$

To summarize, optimal V2G strategies can be derived by the following 3 steps:

- The status of grid-connected EVs are collected, including grid-connected time, expected departing time, the GEV battery SoC state when connecting to the grid, and the preset minimal SoC value at departure.
- The prediction of FA-CD information is carried out based on the method mentioned in Chap. 7. The prediction result is used as the boundary of schedulable charge/discharge power of future arriving EVs. In this way, the solution will have the potential to be global optimal in the whole time domain, and the optimization performance can be improved effectively.
- The optimization cost functions are formulated. With the guidance of cost functions, the V2G strategy could provide grid power balancing, voltage regulation, frequency regulation services, as well as reduce battery degradation.

8.4 Improving V2G Scheduling Performance with Advanced Intelligent Algorithms

The optimization model presented in the previous section can coordinate EV charge/discharge behavior effectively, but it is non-negligible that the V2G coordination is a large-scale, multi-objective problem with $(n + 1) \times (T_u + T_w)$ dimension optimization variable. Thus, when the fleet includes 30 vehicles, the magnitude of optimization variables reaches 10^3. Moreover, the objective to minimize grid load fluctuation somehow conflicts to minimize battery degradation, and the trade-off is actually a problem of searching for a Pareto-optimal point and is normally difficult to solve [45, 46].

PSO is inclined to be stuck in a local optimum during searching because of early maturing, and the evolution process may stop before acquiring the actual global optimum. Particles chase the current global optimum and personal local optimum in memory, while the local optimum gets closer to the global optimum as the swarm iterates and updates [47]. So, particle velocity is soon close to zero, and the whole swarm may accumulate in a small district of search-space, and the search range is limited significantly by this kind of group convergence. To expand the search range, one way is to expand the population size, and the other is to weaken the attraction of global best solution [48, 49]. The former would tremendously increase the computation complexity, and the latter may cause convergence difficulty.

Although PSO has been widely used and validated in single-objective optimization, it is not able to solve multi-objective optimization directly. The key obstacle is also the group homoplasy tendency, which limits much of the search space and deprives the potential to find a coordinating optimal solution. For example, when there are two objectives A and B in optimization, the swarm may gradually accumulate in the district where A gets optimal and lose the tendency to search in B optimal

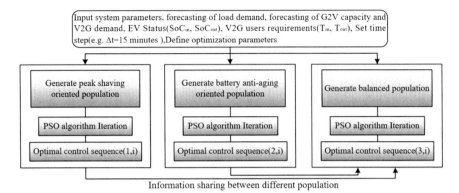

Fig. 8.2 Flow charts of multi-population collaborative particle swarm optimization algorithm for V2G behavior management

district. Then it may be downsized into a single-objective problem, and the output may be inappropriate. This phenomenon is especially obvious when the optimization objectives conflict with each other or the weight of each objective's fitness function is not properly assigned.

Both high-dimensional and multi-objective optimization problems exist in the intelligent V2G management issue. One commonly used method in the existing literature to trickle this problem is to utilize the multi-core, multithreading techniques of modern computers and make better use of parallel computing resources [50, 51]. Multi-population collaborative particle swarm optimization (MC-PSO) algorithm has been proved effective in V2G behavior management in previous literature, which can solve V2G behavior optimization problem with the following 5 steps as shown in Fig. 8.2 [27]:

- Step 1. The status of the grid-connected EVs are collected, including a serial number i of each EV, accessing time t_{start}^i, expected departing time t_{end}^i, SoC_i^{start} when EV_i accesses the grid, and preset minimal SoC_i^{set} at departure. Prediction on FA-CD power boundary $P_{t,discharge}^{pre,\max}$ and $P_{t,charge}^{pre,\max}$.
- Step 2. To satisfy the diversity need in multi-objective optimization, the initial population in the MC-PSO algorithm is divided into three groups: peak-shaving oriented population, battery anti-aging oriented population, and balanced population. Different targets are emphasized when generating initial particles for each population.
- Step 3. Three different populations are independent evolved in the PSO algorithm iteration process. The balanced population represents the coordination between several optimization objectives, and the sub-population can enrich the diversity of the balanced population.
- Step 4. Particles (optimal V2G control sequence) in peak shaving oriented and anti-aging oriented population exchange with balanced population regularly to enrich the diversity of strategy base.

With the above operation mechanism, the quality of the derived V2G strategy can be significantly improved by the information-sharing mechanism between different populations.

8.5 Utilization of GEVs and Grid State Prediction Information in V2G Scheduling

In existing complex system decision theory, predictive control is one of the most commonly used methods to utilize the predicted system state information to improve system performance [52, 53]. Committed to improving the stability and resiliency of the grid, the predicted GEVs energy storage capacity and grid state prediction information should be reasonably utilized in V2G scheduling [54]. The rolling prediction-decision framework, which has been proved effective in V2G behavior management, is introduced in this part to cover the gap between the optimization and forecasting phases in V2G scheduling. As shown in Fig. 8.3, V2G scheduling system is divided into 2 parts: the prediction module and the optimization module in the rolling prediction-decision framework [55]. The optimal V2G strategies are scheduled through the cooperation between the two modules:

- The prediction module predicts the future V2G schedulable capacity and baseload through historical load profile and historical V2G information based on GEVs mobility estimation model presented in Chap. 7. The prediction results are used as the basic load data and V2G schedulable power boundary in the optimization module.
- The V2G behavior of aggregate EVs is coordinated in the optimization module. To downsize the scale of the V2G scheduling problem, the whole V2G scheduling period is divided into two segments: past and future. The global optimization problem is divided into several sub-problem: in each optimization, the V2G

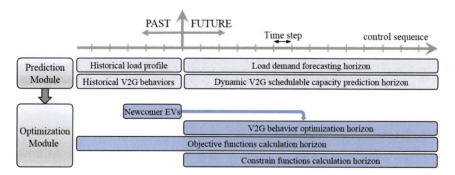

Fig. 8.3 The rolling prediction-decision framework for V2G scheduling

behavior optimization horizon is only the future scheduling period, and the scheduled objects are only the EVs that have just been connected to the grid but not all grid-connected EVs. The V2G behaviors in the past are unmodifiable but influence future strategies, so the objective functions are still calculated in both the future and past periods. The V2G scheduling can be regarded as an optimization problem, and the optimization algorithm is used in this step to derive the optimal V2G strategy.

- The accuracy of the prediction module influences the performance of V2G scheduling directly, so the dynamic prediction control principle is adopted: In the prediction module, to avoid error accumulation in multistep prediction, the historical information is updated along with system operation, and the updated historical data continuously improve the accuracy of the predictive model; Then, in optimization module, the updated predicted information is used to re-optimize the V2G behavior of the newcomer EVs, as described in Step 2. The prediction-decision process is carried out repeatedly with the system operation.

8.6 Strategy Real-Time Deployment: Fuzzy Logic and Machine Learning Method

Based on the optimization model presented in the previous part, the optimal V2G strategy can satisfy the power balancing requirement of the power system and battery anti-aging requirement of participants [56]. However, for the same reason, the real-time performance of the optimization-based model is usually not satisfactory and thus can hardly be used in hardware in the real world [57, 58]. Rule-extraction is one of the most commonly used methods to deploy the strategies from the optimization model [59, 60]. In [61], the energy management strategy of electric vehicles derived by a dynamic programming algorithm is deployed in vehicle controllers by a rule-extraction model. Experimental results indicated that the recalibrated rule-based model could manage the online operation of vehicle power systems while achieving a similar comprehensive performance compared to the dynamic programming model. Combined with the rule extraction method, the optimal V2G strategies derived in the optimization-based model can be deployed in charging devices in the real world. Figure 8.4 shows the procedure for establishing a rule-extraction based V2G behavior management model [62], which can be divided into three steps:

- Firstly, based on the power system's daily operation state information, including grid frequency state, load state and EVs state information, the optimal V2G control strategies are derived by the large-scale optimization algorithm introduced in the previous section. In the optimization process, the grid's peak-shaving and power balancing requirements and the battery anti-aging function can be all considered.
- Then, a rule-based strategy library is established by extracting the characteristics in the derived V2G control strategy. The rules in the strategy library are updated

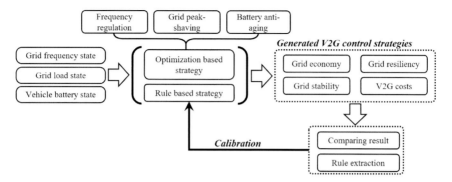

Fig. 8.4 A rule extraction method for real-time V2G strategy deployment

dynamically by comparing the scheduling results of the rule-based controller and optimization method.
- At last, the established rule-based controller is used to directly manage the charging behavior of grid-connected EVs to substitute the optimization-based V2G controller.

In addition, recently developed artificial intelligence algorithms provide a new solution for real-time deploying optimization-based V2G management strategies [63–65]. The optimization and behavior learning method, which uses artificial intelligence to real-time deploy the optimal strategies derived by the optimization process, has been proven to be effective in the industry. In [66], the neural network algorithm is used to deploy energy management strategies derived from the dynamic programming model for plug-in hybrid EVs. The online intelligent energy management controller is established by learning rules from offline scheduling results, and simulation results indicated that the built online controller could accurately reproduce the optimal strategies while guaranteeing real-time performance. In V2G scheduling, the behavior learning methods can also be used to deploy offline strategies. As shown in Fig. 8.5, similar to the fuzzy logic method, real-time V2G strategy deployment consists of the following three steps:

- Firstly, optimal V2G strategy that can satisfy peak-shaving and power balancing requirements of the grid and battery anti-aging requirement of participants are derived from the optimization-based model.
- Then, a neural network-based V2G power controller is established and trained by the optimal strategy derived in the optimization model. The training input is the grid and GEVs states, and the training output is the corresponding V2G power command for each GEV.
- The trained neural network is used to online schedule V2G strategies for participants directly based on the real-time sampled GEVs and grid states information. Because no complex optimization process is used, the real-time performance of the established controller can be effectively guaranteed.

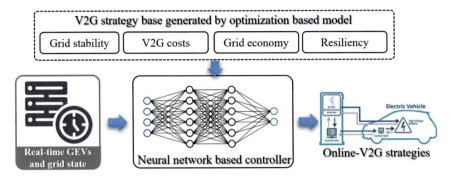

Fig. 8.5 Neural network method for real-time V2G strategy deployment

8.7 Case Study

In this chapter, the real data of load demand of a community with 30 households is used to study the effectiveness of intelligent V2G scheduling. The two most basic V2G management targets: peak-shaving service and vehicle battery aging mitigating, are considered. The most active V2G period that 16:00–24:00 and 00:00–08:00 are taken into consideration. In a random charging scenario, it is assumed that EV owners would immediately charge their cars upon arriving home with rated power until the batteries were fully charged. As shown in Fig. 8.6, most EVs are connected to the grid during 19:00–22:00, while the baseload also increases in this period and peaks at around 21:00, elevating the grid load peak to 504kw. While after 00:00, as most users rest and most EVs are fully charged, there is a valley in the grid load profile, and the minimum grid load is only 100kw. To ensure safe, stable and economic grid operation, it is necessary to suppress the total power imbalance. This can be realized by shifting EV charging load and feeding power back to the grid when necessary.

Based on the V2G scheduling model introduced in part 3, a coordinated charge/discharge scheme is designed, and the results are shown in Fig. 8.7. During

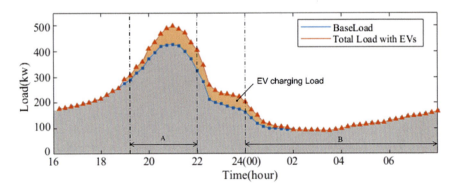

Fig. 8.6 The baseload and the total load profile when EV random charging

grid peak hours, the EV batteries participate in the power exchange, hence no longer overlapping the baseload but feeding power back to the grid. As a result, the peak load is lowered. However, as the PSO algorithm performs poorly on a large-scale optimization problem, its solution is not globally optimal. It can only realize long-term load-shifting but not sufficiently suppress grid load fluctuation, which can be seen in Fig. 8.7 that the load has been fluctuating from 22:00 to 06:00.

The effectiveness of including battery aging mitigation target in V2G scheduling is shown in Fig. 8.8. Subfigure (a) and subfigure (c) present the result of the V2G scheme of an EV without an anti-aging target, and subfigures (b) and (d) present that with anti-aging optimization target. Subfigures (a) and (b) are the SoC profile of an EV from the connection time to the departure time, subfigures (c) and (d) are the charge/discharge cycles statistics based on the rain-flow counting algorithm. The amounts of charge/discharge cycles can be effectively restricted with the battery life loss model and anti-aging optimization target. Half cycles drop from 4 to 3, and full cycles drop more, from 4 to 2. The consideration of battery active anti-aging target can protect vehicle batteries by reducing the amounts of cycles, which highlights the necessity of considering battery aging cost in V2G scheduling.

The grid load profile in the MC-PSO-based V2G scheduling method is shown in Fig. 8.9, which can realize long-term load-shifting and suppress short-term grid load fluctuation. Compared with the random charging scenario, the load peak and load Standard Deviation (STD) is reduced by 23.2% and 41.3%, respectively. While comparing with the conventional PSO algorithm, the load STD is reduced by 20%, but its peak-shaving performance is not satisfying, with only a 1.3% drop compared to baseload. The reason is that the capacity of future arriving EVs is not considered, and the scheme formulated by the aforementioned V2G management system cannot get the global optimal in the whole time domain.

Under the above mechanism, the error accumulation in the prediction process can be avoided by re-optimizing the V2G behaviour of grid-connected EVs, and the effectiveness of the constraints and objectives in V2G scheduling can be enhanced by downsizing the optimization scale. The effectiveness of the rolling time-domain (RTD) method is shown in Fig. 8.10. Compared to the random charging scenario, the load peak and STD are reduced by 31.9% and 60.4%, respectively; compared with

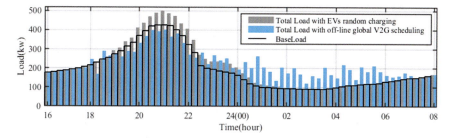

Fig. 8.7 System load profile under multi-objective V2G behavior management method

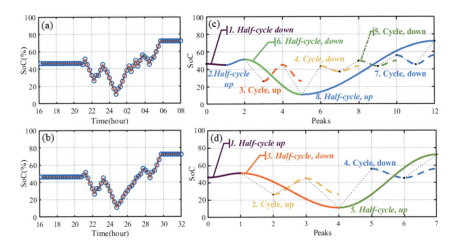

Fig. 8.8 Comparing battery aging cycles in V2G scheduling methods with (**a**, **b**) and without (**c**, **d**) battery anti-aging target

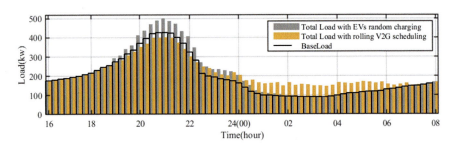

Fig. 8.9 System load profile under MC-PSO algorithm

the V2G management scheme based on the MC-PSO algorithm, the load peak and STD can be further reduced by 11.4% and 32.6% respectively.

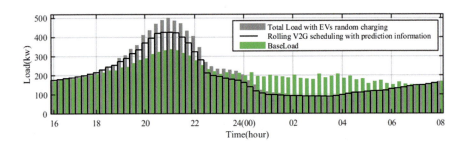

Fig. 8.10 System load profile under rolling time-domain method

8.8 Conclusion

This chapter performed a thorough literature review of state-of-the-art research works in terms of V2G behaviour management and optimization. By introducing the multi-objective GEVs charging management model, V2G benefits in terms of grid peak-shaving and battery anti-aging are realized. Afterwards, the MC-PSO optimization algorithm is introduced, one of the most proper algorithms for solving V2G management problems that can reasonably utilize the computational resources. Furthermore, the prediction control method is also introduced for enhancing the power balancing performance of V2G scheduling by better-utilizing prediction information. In addition, the two most commonly used real-time strategy deployment methods: fuzzy logic and neural network, are introduced for online V2G behavior management. To illustrate the effectiveness of V2G scheduling, grid load demand of a community with 30 households 30 GEVs are used to simulate power system operation. Results show that grid load peak-valley difference can be significantly reduced through coordinating the charging behavior of GEVs. Furthermore, simulation results also indicate the necessity of deploying anti-aging targets and using prediction information in V2G scheduling systems, which can notably improve power balancing performance and vehicle battery aging costs.

References

1. D.T. Hoang, P. Wang, D. Niyato, E. Hossain, Charging and discharging of plug-in electric vehicles (PEVs) in vehicle-to-grid (V2G) systems: a cyber insurance-based model. IEEE Access **5**, 732–754 (2017)
2. Q. Zhang, Y. Zhu, Z. Wang, Y. Su, C. Li, Reliability assessment of distribution network and electric vehicle considering quasi-dynamic traffic flow and vehicle-to-grid. IEEE Access **7**, 131201–131213 (2019)
3. P.-Y. Kong, G.K. Karagiannidis, Charging schemes for plug-in hybrid electric vehicles in smart grid: a survey. IEEE Access **4**, 6846–6875 (2016)
4. Y. Xu, Optimal distributed charging rate control of plug-in electric vehicles for demand management. IEEE Trans. Power Syst. **30**(3), 1536–1545 (2015)
5. Y. Ota, H. Taniguchi, T. Nakajima, K.M. Liyanage, J. Baba, A. Yokoyama, Autonomous distributed V2G (Vehicle-to-Grid) satisfying scheduled charging. IEEE Trans. Smart Grid **3**(1), 559–564 (2012)
6. C.-K. Wen, J.-C. Chen, J.-H. Teng, P. Ting, Decentralized plug-in electric vehicle charging selection algorithm in power systems. IEEE Trans. Smart Grid **3**(4), 1779–1789 (2012)
7. O. Rahbari et al., An optimal versatile control approach for plug-in electric vehicles to integrate renewable energy sources and smart grids. Energy **134**, 1053–1067 (2017)
8. S. Vachirasricirikul, I. Ngamroo, Robust LFC in a smart grid with wind power penetration by coordinated V2G control and frequency controller. IEEE Trans. Smart Grid **5**(1), 371–380 (2014)
9. M. Singh, P. Kumar, I. Kar, Implementation of vehicle to grid infrastructure using fuzzy logic controller. IEEE Trans. on Smart Grid **3**(1), 565–577 (2012)
10. K. Clement-Nyns, E. Haesen, J. Driesen, The impact of charging plug-in hybrid electric vehicles on a residential distribution grid. IEEE Trans. Power Syst. **25**(1), 371–380 (2010)

11. M.R. Sarker, Y. Dvorkin, M.A. Ortega-Vazquez, Optimal participation of an electric vehicle aggregator in day-ahead energy and reserve markets. IEEE Trans. Power Syst **31**(5), 3506–3515 (2016)
12. Z. Hu, et al., Multi-objective energy management optimization and parameter sizing for proton exchange membrane hybrid fuel cell vehicles. Energy Convers. Manag. **129**, 108–121 (2016)
13. H. Cai, W. Du, X. P. Yu, S. Gao, T. Littler, H.F. Wang, Day-ahead optimal charging/discharging scheduling for electric vehicles in micro-grids, in *2nd IET Renewable Power Generation Conference (RPG 2013)* (2013), pp. 1–4
14. A. Dogan, S. Bahceci, F. Daldaban, M. Alci, Optimization of charge/discharge coordination to satisfy network requirements using heuristic algorithms in vehicle-to-grid concept. Adv. Electr. Comput. Eng. **18**(1), 121–131 (2018)
15. Y. He, B. Venkatesh, L. Guan, Optimal scheduling for charging and discharging of electric vehicles. IEEE Trans. Smart Grid **3**(3), 1095–1105 (2012)
16. L. Gan, U. Topcu, S.H. Low, Optimal decentralized protocol for electric vehicle charging. IEEE Trans. Power Syst. **28**(2), 940–951 (2013)
17. H. Wu, G.K.-H. Pang, K.L. Choy, H.Y. Lam, Dynamic resource allocation for parking lot electric vehicle recharging using heuristic fuzzy particle swarm optimization algorithm. Appl. Soft Comput. **71**, 538–552 (2018)
18. O. Kolawole, I. Al-Anbagi, Electric vehicles battery wear cost optimization for frequency regulation support. IEEE Access **7**, 130388–130398 (2019)
19. T. Mao, B. Zhou, X. Zhang, Accommodating discharging power with consideration of both EVs and ESs as commodity based on a two-level GA algorithm. IEEE Access (2019)
20. S. Li, C. Gu, P. Zhao, S. Cheng, Adaptive energy management for hybrid power system considering fuel economy and battery longevity. Energy Convers. Manag. **235**, 114004 (2021)
21. C. Li et al., An optimal coordinated method for EVs participating in frequency regulation under different power system operation states. Ieee Access **6**, 62756–62765 (2018)
22. A. Ahmadian, M. Sedghi, B. Mohammadi-ivatloo, A. Elkamel, M.A. Golkar, M. Fowler, Cost-benefit analysis of V2G implementation in distribution networks considering PEVs battery degradation. IEEE Trans. Sustain. Energy **9**(2), 961–970 (2017)
23. M. Landi, G. Gross, Measurement techniques for online battery state of health estimation in vehicle-to-grid applications. IEEE Trans. Instrum. Meas. **63**(5), 1224–1234 (2014)
24. S. Tabatabaee, S.S. Mortazavi, T. Niknam, Stochastic scheduling of local distribution systems considering high penetration of plug-in electric vehicles and renewable energy sources. Energy **121**, 480–490 (2017)
25. T.S. Ustun, C.R. Ozansoy, A. Zayegh, "Implementing vehicle-to-grid (V2G) technology with IEC 61850-7-420. IEEE Trans. Smart Grid **4**(2), 1180–1187 (2013)
26. C. Guille, G. Gross, A conceptual framework for the vehicle-to-grid (V2G) implementation. Energy Policy **37**(11), 4379–4390 (2009)
27. S. Li, J. Li, C. Su, Q. Yang, Optimization of bi-directional V2G behavior with active battery anti-aging scheduling. IEEE Access **8**, 11186–11196 (2020)
28. A.M. Ghazvini, J. Olamaei, Optimal sizing of autonomous hybrid PV system with consider-ations for V2G parking lot as controllable load based on a heuristic optimization algorithm. Solar Energy **184**, 30–39 (2019)
29. H.N. Nguyen, C. Zhang, M.A. Mahmud, Optimal coordination of G2V and V2G to support power grids with high penetration of renewable energy. IEEE Trans. Transp. Electrif. **1**(2), 188–195 (2015)
30. X. Wang, Y. Nie, K.-W.E. Cheng, Distribution system planning considering stochastic EV penetration and V2G behavior. IEEE Trans. Intell. Transp. Syst. **21**(1), 149–158 (2019)
31. Z. Moghaddam, I. Ahmad, D. Habibi, M.A. Masoum, A coordinated dynamic pricing model for electric vehicle charging stations. IEEE Trans. Transp. Electrif. **5**(1), 226–238 (2019)
32. M.-H. Khooban, T. Niknam, M. Shasadeghi, T. Dragicevic, F. Blaabjerg, Load frequency control in microgrids based on a stochastic noninteger controller. IEEE Trans. Sustain. Energy **9**(2), 853–861 (2017)

33. E. Hazan, Introduction to online convex optimization. Foundations Trends® in Optimization, vol. 2, no. 3–4, pp. 157–325 (2016)
34. S. Ruder, An overview of gradient descent optimization algorithms. arXiv:.04747 (2016)
35. T.T. Mac, C. Copot, D.T. Tran, R. De Keyser, A hierarchical global path planning approach for mobile robots based on multi-objective particle swarm optimization. Appl. Soft Comput. **59**, 68–76 (2017)
36. B.P. De, R. Kar, D. Mandal, S.P. Ghoshal, Optimal selection of components value for analog active filter design using simplex particle swarm optimization. Int. J. Mach. Learn. Cybern. **6**(4), 621–636 (2015)
37. M.A. Rahman, S. Anwar, A. Izadian, Electrochemical model parameter identification of a lithium-ion battery using particle swarm optimization method. J. Power Sour. **307**, 86–97 (2016)
38. J. Yang, L. He, S. Fu, An improved PSO-based charging strategy of electric vehicles in electrical distribution grid. Appl. Energy **128**, 82–92 (2014)s
39. S. Hajforoosh, M.A. Masoum, S.M. Islam, Real-time charging coordination of plug-in electric vehicles based on hybrid fuzzy discrete particle swarm optimization. Electr. Power Syst. Res. **128**, 19–29 (2015)
40. J. Soares, H. Morais, T. Sousa, Z. Vale, P. Faria, Day-ahead resource scheduling including demand response for electric vehicles. IEEE Trans. Smart Grid **4**(1), 596–605 (2013)
41. S. Li, C. Gu, M. Xu, J. Li, P. Zhao, S. Cheng, Optimal power system design and energy management for more electric aircrafts. J. Power Sour. **512**, 230473 (2021)
42. J. Hu, C. Ye, Y. Ding, J. Tang, S. Liu, A distributed MPC to exploit reactive power V2G for real-time voltage regulation in distribution networks. IEEE Trans. Smart Grid **13**(1), 576–588 (2021)
43. Y. Huang, Day-ahead optimal control of PEV battery storage devices taking into account the voltage regulation of the residential power grid. IEEE Trans. Power Syst. **34**(6), 4154–4167 (2019)
44. F. Kennel, D. Görges, S. Liu, Energy management for smart grids with electric vehicles based on hierarchical MPC. IEEE Trans. Industr. Inf. **9**(3), 1528–1537 (2013)
45. V. Beiranvand, M. Mobasher-Kashani, A.A. Bakar, Multi-objective PSO algorithm for mining numerical association rules without a priori discretization. Expert Syst. Appl. **41**(9), 4259–4273 (2014)
46. J. Wu, C.-H. Zhang, N.-X. Cui, PSO algorithm-based parameter optimization for HEV powertrain and its control strategy. Int. J. Automot. Technol. **9**(1), 53–59 (2008)
47. W. Deng, R. Yao, H. Zhao, X. Yang, G. Li, A novel intelligent diagnosis method using optimal LS-SVM with improved PSO algorithm. Soft Comput. **23**(7), 2445–2462 (2019)
48. N. B. J. A. S. C. Guedria, "Improved accelerated PSO algorithm for mechanical engineering optimization problems," vol. 40, pp. 455–467, 2016.
49. W. Deng, H. Zhao, X. Yang, J. Xiong, M. Sun, B. Li, Study on an improved adaptive PSO algorithm for solving multi-objective gate assignment. Appl. Soft Comput. **59**, 288–302 (2017)
50. S. Abdi, S.A. Motamedi, S. Sharifian, Task scheduling using modified PSO algorithm in cloud computing environment, in *International Conference on Machine Learning, Electrical and Mechanical Engineering*, vol. 4, no. 1 (2014), pp. 8–12
51. J.-Y. Kim, K.-J. Mun, H.-S. Kim, J.H. Park, Optimal power system operation using parallel processing system and PSO algorithm. Int. J. Electr. Power Energy Syst. **33**(8), 1457–1461 (2011)
52. Y. Wang, Z. Wang, L. Zhang, M. Liu, J. Zhu, Lateral stability enhancement based on a novel sliding mode prediction control for a four-wheel-independently actuated electric vehicle. IET Intell. Trans. Syst. **13**(1), 124–133 (2019)
53. Y. Peng, A. Rysanek, Z. Nagy, A. Schlüter, Using machine learning techniques for occupancy-prediction-based cooling control in office buildings. Appl. Energy **211**, 1343–1358 (2018)
54. A. Haque, V.S.B. Kurukuru, M.A. Khan, Stochastic methods for prediction of charging and discharging power of electric vehicles in vehicle-to-grid environment. IET Power Electr. **12**(13), 3510–3520 (2019)

55. S. Li, C. Gu, J. Li, H. Wang, Q. Yang, Boosting grid efficiency and resiliency by releasing V2G potentiality through a novel rolling prediction-decision framework and deep-LSTM algorithm. IEEE Syst. J. **15**(2), 2562–2570 (2021)
56. K. Thirugnanam, E.R.J. TP, M. Singh, P. Kumar, Mathematical modeling of Li-ion battery using genetic algorithm approach for V2G applications. IEEE Trans. Energy Convers. **29**(2), 332–343 (2014)
57. A. Zakariazadeh, S. Jadid, P. Siano, Multi-objective scheduling of electric vehicles in smart distribution system. Energy Convers. Manag. **79**, 43–53 (2014)
58. S. Li, C. Gu, X. Zeng, P. Zhao, X. Pei, S. Cheng, Vehicle-to-grid management for multi-time scale grid power balancing. Energy **234**, 121201 (2021)
59. J. Wang, J. Wang, Q. Wang, X. Zeng, Control rules extraction and parameters optimization of energy management for bus series-parallel AMT hybrid powertrain. J. Frankl. Inst. **355**(5), 2283–2312 (2018)
60. P. May-Ostendorp, G. P. Henze, C. D. Corbin, B. Rajagopalan, C. Felsmann, Model-predictive control of mixed-mode buildings with rule extraction. Build. Envir. **46**(2), 428–437 (2011)
61. J. Peng, H. He, R. Xiong, Rule based energy management strategy for a series–parallel plug-in hybrid electric bus optimized by dynamic programming. Appl. Energy **185**, 1633–1643 (2017)
62. S. Li, et al., Online battery-protective vehicle to grid behavior management. Energy **243**, 123083 (2022)
63. Y. Liu, et al., Prediction of vehicle driving conditions with incorporation of stochastic forecasting and machine learning and a case study in energy management of plug-in hybrid electric vehicles. Mech. Syst. Signal Process. **158**, 107765 (2021)
64. X. Sun and J. Qiu, A customized voltage control strategy for electric vehicles in distribution networks with reinforcement learning method. IEEE Trans. Ind. Inf. **17**(10), 6852–6863 (2021)
65. K.M. Tan, S. Padmanaban, J.Y. Yong, V.K. Ramachandaramurthy, A multi-control vehicle-to-grid charger with bi-directional active and reactive power capabilities for power grid support. Energy **171**, 1150–1163 (2019)
66. Z. Chen, C.C. Mi, J. Xu, X. Gong, C. You, Energy management for a power-split plug-in hybrid electric vehicle based on dynamic programming and neural networks. IEEE Trans. Veh. Technol. **63**(4), 1567–1580 (2014)

Chapter 9
Local Energy Trading with EV Flexibility

Shuang Cheng, Da Xie, and Chenghong Gu

Abstract The rapid development of electric vehicles (EVs) brings challenges and opportunities for electricity systems. EV flexibility, such as smart charging and vehicle-to-grid services, plays a crucial role in helping systems to integrate renewable generation, thus facilitating achieving Net-Zero carbon targets. Market measures are of great help to incorporate a sufficient level of flexibility for electricity system operation requirements. Specifically, demand-side flexibility (DSF) should be fairly rewarded when providing various services to the systems. EV flexibility trading effectively supports the integration of renewable energy resources (RERs) and uncertain demand. Local energy trading mechanisms with EV flexibility can help efficiently facilitate flexibility markets. This chapter classifies EV flexibility and investigates the potential values for the electricity system. The local market structures and trading mechanisms for enabling EV flexibility are introduced. An example demonstration shows that integrating EV flexibility in the local energy market allows for more efficient operation.

Keywords Electric vehicle · Smart charging · Vehicle-to-grid · Flexibility market · Net zero · Demand-side flexibility · Renewable energy resources

Abbreviations

DSF	Demand-side Flexibility
EV	Electric Vehicle
P2P	Peer-to-Peer
HEV	Hybrid EV
FCEV	Fuel Cell EV

S. Cheng · C. Gu (✉)
Department of Electronic and Electrical Engineering, Universiy of Bath, Bath, UK
e-mail: cg277@bath.ac.uk

D. Xie
Department of Electrical Engineering, Shanghai Jiao Tong University, Shanghai, China

© The Author(s), under exclusive license to Springer Nature Singapore Pte Ltd. 2023
Y. Cao et al. (eds.), *Automated and Electric Vehicle: Design, Informatics and Sustainability*, Recent Advancements in Connected Autonomous Vehicle Technologies 3, https://doi.org/10.1007/978-981-19-5751-2_9

PETCON	P2P Electricity Trading System with Consortium Blockchain
PV	Photovoltaic
DA	Day-ahead
FIAD	Flexibility Index of Aggregate Demand
PFL	Percentage Flexibility Level
BEV	Battery EV
PHEV	Plug-in Hybrid EV
DSO	Distribution System Operator
CCGT	Combined Cycle Gas Turbine
WT	Wind Turbine
MES	Multi-Energy System

9.1 Introduction

Many countries have set ambitious emission reduction targets to tackle climate change. The UK government sets carbon emission reductions by 78% by 2035 compared to 1990 to achieve net-zero by 2050. China has made a concerted effort to hit the peak CO_2 emissions before 2030 and reach carbon neutrality by 2060. The US set a target to reduce carbon emissions by 50–52% below 2005 levels by 2030. Decarbonization technologies are indispensable to meet the requirements of ambitus carbon targets, i.e., heat and transport electrification, renewable energy resources (RERs), etc. With the growing shares of intermittent and uncertain renewable generation, the power system is increasingly vulnerable to unexpected operation conditions and capacity shortages [1]. On the other hand, extensive electrification creates new sources of flexibility, for example, electric vehicles, batteries, and smart demand. It is possible to aggregate and sell demand-side flexibility (DSF) explicitly. DSF also provides alternative operation measures to manage distribution congestion. In conclusion, the high integration of RERs calls for new sources of flexibility, and transport electrification, e.g., electric vehicles (EVs), increases the potential to provide demand-side flexibility [2]. The key challenges are to unlock DSF through smart energy products and services to improve RER shares, thus avoiding considerable system costs and achieving environmental targets.

The key barriers that need urgent solutions in promoting EV flexibility are as follows:

(1) Policy and market reform. Pricing methods and market reform measures [3–5] should be designed to reflect the full value of various types of EVs. Appropriate reform measures enable easier access for EV flexibility to all markets, e.g., the day-ahead wholesale market, the balancing market, and the capacity market. Pricing frameworks should provide regulatory economic incentives for market participants to support EV flexibility. Potential research orientation includes (i) access and forward-looking charging methods; (ii) ancillary services; (iii) balancing mechanism; (iv) wholesale market.

(2) Business models. To facilitate a smart and flexible energy system, effective business models [6] should be formulated to coordinate local and national markets to enable all levels from domestic, community energy, and non-domestic customers. For instance, they should support effective EV aggregation to incentivize consumer participation and minimize system operating costs.

(3) New technologies. Supportive infrastructure should be deployed to enable accessible EV flexibility, e.g., smart meters and EV charge point standards. For instance, in conjunction with communication systems, smart metering is essential to monitor real-time energy consumption, which provides valuable tools to encourage flexible customers to participate in energy markets. In addition, numerous mathematical models are being developed to support EV flexibility in the energy market, i.e., blockchain and game theory [7, 8].

Local energy markets, e.g., local transactive energy market and peer-to-peer (P2P) market [6], accommodate promising solutions to access and stimulate EV flexibility in energy systems. These solutions make it possible to trade and balance local energy between prosumers at the distribution level. Since DSF will play a significant role in the future local market, corresponding trading mechanisms are needed to overcome the barriers and challenges.

9.2 The Flexibility of Electric Vehicles

9.2.1 Electricity System Flexibility

The term—flexibility is used to indicate the power system's capacity to adapt across time, circumstances, intention, and area of application [9]. In the electrical system, the definition has been discussed in many research studies. In [10], system flexibility is defined as *"the ability of a power system to cope with variability and uncertainty in both generation and demand, while maintaining a satisfactory level of reliability at a reasonable cost, over different time horizons"*. In [11], flexibility is described as *"the ability of a system to deploy its resources to respond to changes in net load, where the net load is defined as the remaining system load not served by variable generation"*. Reference [12] presents that flexibility refers to *"the possibility of deploying the available resources to respond in an adequate and reliable way to the load and generation variations during the time at acceptable costs"*. Moreover, DSF is mathematically defined in [12] for aggregate residential loads based on the probability of changing their collective behavior in different periods. It proposes two flexibility indicators: i.e. (1) flexibility index of aggregate demand (FIAD), which denotes the probability of demand increase and decrease; (2) percentage flexibility level (PFL), which denotes the percent amount of flexible demand available for demand-side management measures. These flexibility evaluation indexes enable system operators to formulate demand response programs in terms of selecting available time slots.

9.2.2 EV Flexibility

EVs can be generally divided into four types: battery EV (BEV), hybrid EV (HEV), plug-in hybrid EV (PHEV), and fuel cell EV (FCEV). BEV stores the electricity from the grid in batteries to power the electric motor. Known as a series hybrid vehicle, PHEV uses electricity and liquid fuel (e.g., gasoline) to drive the electric motor and the internal combustion engine, respectively [13]. HEV is also called a standard hybrid or parallel hybrid vehicle with a similar structure to PHEV. Its battery can be charged alternatively by the internal combustion engine, wheel motions, or the combination of both, which distinguishes it from PHEV and BEV.

Although transport electrification will increase overall electricity demand, it also enables flexibility potentials for the electricity system. EV smart energy management (e.g., smart vehicle-to-grid charging) can provide system operators with considerable flexibility. Compared to conventional vehicles, EV has different energy storage systems that can provide flexibility that is affected by storage types and operation mechanisms.

Diverse EV charging methods lead to a variety of flexibility. The charging locations and speed are the two important factors for different EVs. In general, there are three charging categories, i.e., trickle charge, AC charge, and DC charge [15]. Trickle charge is the slowest method of charging vehicles at home using a 220 V plug. AC charging is available for both households and public charging, with a deployed wall box that is 3–4 faster than trickle charge. DC charging is the fastest approach with power exceeding 50 kW. It enables the battery to be charged from 20 to 80% in around 40 min [15]. Nevertheless, it is commonly restricted to public charging stations, where EV customers do not have much time to recharge. The big differences between these charging methods are summarized in Table 9.1.

EV flexibility plays a significant role in incorporating more variability and uncertainty into electricity systems, i.e., energy supply uncertainty and demand uncertainty. Figure 9.1 demonstrates the contribution of EV flexibility to the electricity system under an operation cost constraint in a power system. The smaller blue circuit represents a power system without EV flexibility capacity, while the larger green circuit demonstrates that with flexible EVs. The horizontal and vertical axes denote demand

Table 9.1 Differences of major charging methods

Methods	Trickle charge	AC charge with wallbox	DC fast charge
Charging station	Household	Household Public stations	Public stations
Charging speed	Approx. 65 km of range in 5 h	Approx. 6 h to fully charge a 40 kWh battery car	Approx. 40 min from 20 to 80% of charge
Scenario	Only recommended in urgent cases	Most common and recommendable home charging option Public charging option	The most common option for public charging

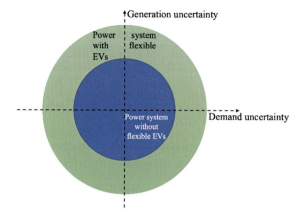

Fig. 9.1 Demonstration of EV flexibility values to absorb uncertainties in a power system

uncertainty and generation uncertainty tolerance, respectively. The figure shows that the green circuit intersects the two axes with higher values, demonstrating that the flexible power system can accommodate more generation and demand uncertainties under a given cost threshold. Therefore, facilitating EV flexibility in the power system is significant to absorbing more RERs and thus achieving carbon targets.

To apply EV flexibility in the electricity system, appropriate quantification tools should be developed to evaluate flexibility's economic and environmental values. Reference [16] proposes a novel EV intelligent integrated station to accommodate the increasing electric vehicles (EVs) and maximize their benefits to the grid. The framework consists of a dispatching center, multiuse converter devices, a charge exchange system, and an echelon battery system. Results demonstrate that EV flexibility can effectively offer peak load shifting through the proposed charging and discharging control strategy. It makes full use of ex-service batteries and thus maximizes the benefits of EVs more flexibly and effectively. Reference [17] investigates the capability of current and potential accessible future residential DSF (e.g., EVs. Stationary batteries, and storage heaters) to reduce the energy supply costs of a flexible aggregator with highly penetrated RERs. Results demonstrate that even a low saturation of flexible demand can reduce generation costs in microgrids. Additionally, by considering PV and wind generation as uncertain renewables, paper [17] finds that flexible loads have considerable value in a PV-dominated microgrid but are of little importance in a wind-dominated microgrid. Particularly, the economic values of stationary batteries are remarkable in the PV-dominated portfolio. Nevertheless, a wind-dominated microgrid has seen the same performance as stationary batteries and EVs.

In terms of the assessment model for EVs, paper [18] quantifies their flexibility and incorporates it in the system planning model. It co-optimizes both the investment and operating costs of EVs and generation assets. Results indicate that EV flexibility plays an important role in reducing peak demand and addressing wind generation uncertainty. Its economic value increases with the growing electrification of the

Table 9.2 Potential EV flexibility benefits in the power system

Stage	EV flexibility benefits
Generation	Generation cost reduction for aggregators in PV or wind dominated generation portfolios [15]
Transmission	• System operation cost curtailment by reducing peak demand levels [16] • Short-term congestion management in the day-ahead and balancing market through EV flexibility services to address demand uncertainty [17] • EV flexibility as an alternative to deferring or even avoiding network reinforcement (e.g., laying cables and building transformers) [18–20]
Distribution	• EV customers can save on energy expenses through incentive tariffs (e.g., time-varying tariffs) • Aggregators can obtain considerable economic benefits through arbitrage behaviors

transport sector and wind generation. Potential flexibility benefits in different stages of the electricity supply chain are summarized in Table 9.2.

9.3 Local Flexibility Market

To facilitate flexibility services, they can participate in the electricity market through central scheduling, decentralized trading, and P2P, as presented below. Through central scheduling, DSF can be sold in wholesale markets, where flexibility bids/offers can be accepted by system operators using the merit order method. Decentralized markets enable distribution system operators, aggregators, and prosumers to negotiate transactions with each other. P2P market supports individual flexibility buyers and sellers to establish bilateral contracts, increasing the market liquidity.

9.3.1 Central Markets

EV flexibility can be utilized in the power system on both the transmission and distribution level through centralized control. The local system services procured from EV flexibility are focused on in this chapter. Figure 9.2 illustrates a typical centralized framework for distribution system operators (DSOs) and flexibility providers. As shown in this figure, there are two markets for ancillary services procurement by DSO from EV flexibility. In the reservation market, EV flexibility aggregators submit bids to and receive availability payment from DSO. The bids in the reservation market include the volumes and reservation time of available flexible resources. Those reserved flexibility providers are responsible for participating in the activation market with new providers. Bids submitted to the activation market include volumes, activated hours, and activation prices of flexible resources. They are paid through the utilization prices.

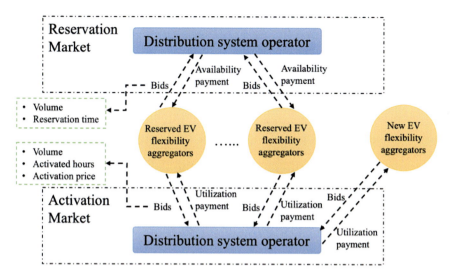

Fig. 9.2 Framework for centralized flexibility market in the distribution-level between distribution system operators and flexibility providers

In terms of the time scale of dual markets, the reservation market can take place from year-ahead to day-ahead [20]. Long-term needs are valuable for system planning, while short-term actions enable a more competitive market environment. The activation market is cleated in real-time based on the common merit-order method. Ex post-settlement is reached through the reservation price and clearing price in the dual markets, respectively.

9.3.2 Decentralized Markets

Considering the small size of individual prosumers, they cannot participate in the wholesale market directly. Instead, they can be aggregated through a third party. The decentralized market enables distribution network operators to procure flexibility from competing aggregators. Aggregators award the prosumers to optimize their demand consumption behavior to provide flexibility. To simultaneously benefit the DSO, aggregators, and prosumers, Pareto efficiency should be satisfied in the designed decentralized market [23].

Based on the existing retail market, a typical framework for the decentralized market is as shown in Fig. 9.3. As seen in the figure, the flexibility is traded between three types of agents, i.e., the DSO, prosumers, and aggregators. The DSO negotiates with aggregators rather than directly with the prosumers or customers for flexibility services through multiple contracts at each time interval. The trading time interval is set as the duration when the retail supplier meters customers' individual net energy use. Correspondingly, the aggregators sign flexibility contracts with prosumers or

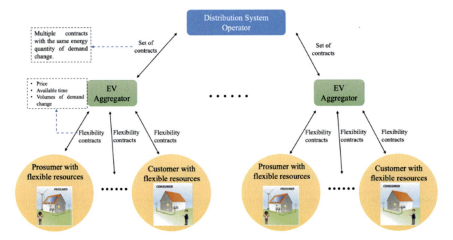

Fig. 9.3 Framework for decentralized flexibility market between distribution system operators, aggregators, and local prosumers

customers that can provide flexible resources. The price, the available scheduling time, and the volumes for demand changes are specified in those contracts. Remarkably, each customer's energy flexibility contract can be priced individually. The market is cleared through an iterative price-negotiation mechanism [23]. Specifically, agents construct their available contracts respectively and thereafter select their preferred ones. For each iteration, the buyer price or the seller price increase by a price unit until they are equal, i.e., the negotiation is complete with unchanged prices.

9.3.3 P2P Markets

To fully explore the potential of market participants and exert the market advantages for flexibility facilitation, bilateral energy transactions can be designed. This P2P configuration allows individual flexibility buyers and sellers to negotiate with each other. Considering most DSF resources are subject to uncertainties due to weather changes or human behaviors. The aforementioned two structures enforce financial penalties in case the providers fail to stick to their submitted volumes and available time [24]. By contrast, P2P trading enables individual prosumers or aggregators to establish bilateral transactions so that those risks from unsuccessful delivery can be hedged.

A typical P2P market structure is shown in Fig. 9.4. The flexibility buyers and sellers can be household end-customers/prosumers who are connected with EVs, storage systems, PV, wind turbines, or flexible household loads. Flexibility bids

9 Local Energy Trading with EV Flexibility

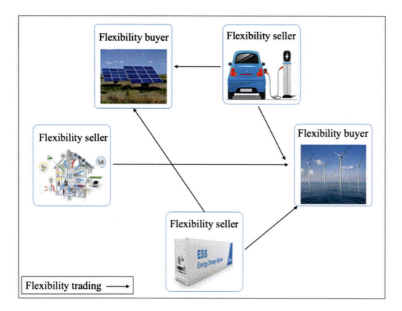

Fig. 9.4 Framework for P2P structure between flexibility sellers and buyers

and offers are submitted to this platform and cleared through multiple decentralized technologies, such as Blockchain or auction mechanisms.

Notably, the choice of the trading structures on real power systems depends on the market development. With relatively undeveloped markets, a hybrid system is more suitable. However, a move to a fully market-driven approach is expected eventually.

9.4 Trading Mechanisms with EV Flexibility

By choosing the appropriate evaluation model and market structure, corresponding trading mechanisms and clearing algorithms can be designed for EV flexibility in the local energy system. For centralized market architecture, the clearing function is to minimize the total payment of DSO to buy flexibility services, which are DSO-oriented. In the settlement process, pay-as-bid (discriminatory) and pay-as-clear (uniform) clearing mechanisms can be used to calculate the cleared price for all providers. The uniform pricing mechanism sets the same clearing price for all participants as the most expensive bid accepted. In comparison, under pay-as-bid auction rules, participants are paid the price that they bid at. The result of this market-clearing procedure allocates the flexibility services oriented towards cost minimization for DNO [20].

Decentralized and P2P market structures call for a different and more complicated clearing mechanism because the price negotiation is bilateral depending on individual decision-making. Lots of decentralized trading approaches are developed to address those price negotiation problems [23]. To capture the stochastic bidding process of EV flexibility providers to buy and sell electricity with their maximized profits, some popular algorithms are proposed. They can be divided into three categories [25]:

- Auction-based model [26, 27]: it enables EV flexibility prosumers to transact with each other through energy biding or offering and auction mechanisms in the local energy market.
- Multi-agent model [28, 29]: it accounts for multiple autonomous and interactive agents with conflicting or common objectives in the local trading system. Depending on the degree of complexity, the models can be designed for multiple agents with three classes: no interaction, simple interaction, and complex interaction [30].
- Analytical model [31–33]: it formulates the pricing mechanism for EVs through particular rules or game-theoretic methods.

9.4.1 Blockchain

Borrowed from finance, blockchain technology is a decentralized network where all participants can transact through shared responsibility rather than central authority to verify transactions [34]. By means of smart contracts, blockchain technology can be applied to flexibility trading for the distribution network or microgrid [35, 36]. It enables more efficient and secure trading platforms, especially with a massive trading database. A consortium blockchain scheme is proposed in [37] to demonstrate the plug-in hybrid EV trading in smart grids. With the aim of overall social welfare maximization, the proposed method is a localized P2P electricity trading system with consortium blockchain (PETCON) that can address security, computation efficiency, and privacy protection problems.

The trading procedures are illustrated in Fig. 9.5. As shown in this figure, there are three entities in a local flexibility aggregator, i.e., account pool, transaction server, and memory pool. The transaction server collects discharging/charging requests from EVs, combines matched sides of the trade, and controls switches of EV charging stations for energy delivery. The account pool archives each EV's wallet address and account information. The memory pool saves all transaction records connected to this local aggregator. In this way, the local aggregator can execute the EV bidding process and complete transactions by a double auction mechanism [38]. Structured as a block, each local aggregator is connected to the prior block through a cryptographic hash in the blockchain.

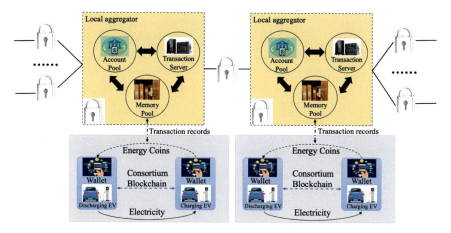

Fig. 9.5 Trading structure of the PETCON system

9.4.2 Game Theory

Considering that end customers with flexible resources have conflicting interests, it is difficult to capture their behaviors in the decision-making process. Furthermore, it is more challenging to incentivize them to cooperate for achieving the system's objectives in the P2P trading process [25]. To model the random decision-making process of competitive participants [25], the game-theoretical approaches are investigated in many studies to ensure both optimization and fairness among EV flexibility providers.

The energy trading problem can be modeled as a single-leader and multiple follower Stackelberg games, as illustrated in [39]. Following the game rules, the DSO is a leader who establishes the pricing scheme to maximize its revenues, while the followers, i.e., prosumers, in turn, respond to it to optimize their benefits through energy transactions. The framework of the Stackelberg game-based trading platform is shown in Fig. 9.6. It can be found in the figure that the utility function of DSO is formulated by considering the current market-clearing price and the energy bids from EV prosumers. The pricing strategy is designed by maximizing the revenues based on the established utility function. In terms of the EV consumers, the prospect theory is used to evaluate each EV's potential gains and losses underprice uncertainty and individual utility reference point. The reference point is derived from their past experience and future expectations of profits. In such a case, EV prosumers' subjective behaviors are captured. The pure-strategy Nash equilibrium under classical game theory is achieved to simulate the noncooperative game of multiple EV prosumers. The overall trading platform is modeled as a hierarchical Stackelberg game with the DSO as a single leader in the upper stage and the EV prosumers as followers in the lower stage. The game-theoretical methods allow for human behaviors in the overall flexibility trading procedure.

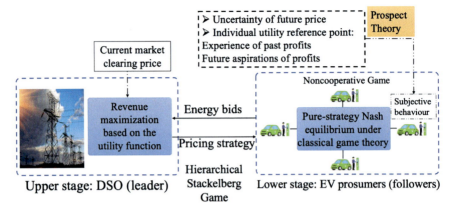

Fig. 9.6 P2P trading platform based on the Stackelberg game for EV prosumers and DSO

9.4.3 Machine Learning

The EV users' behaviors are influenced by many factors (e.g., EV availability), which highly complicates their participation in the local energy market. To fully unlock the potential of EV flexibility trading, key drivers of individual EV decisions [40] should be captured to optimize the overall system's operational efficiency. Therefore, machine learning techniques are utilized to simulate the random behaviors of individual EV owners by learning the key factors driving their decisions. For instance, the Naïve Bayes model can be used to optimize the objective function through Bayesian statistics [41]. Specifically, it can predict the class of test dataset with $T = a_1, a_2, \ldots, a_n$ by selecting the class c_i, to optimize the formulated objective function. The conditional probability of class c_i and $a_j|c_i$ (i.e., $P(c_i)$ and $P(a_j|c_i)$) determine the predictions, which are shown in (1) and (2) (Table 9.3).

$$P(c_i|T) = \frac{P(T, c_i)}{P(T)} = \frac{P(c_i) \cdot P(T|c_i)}{P(T)} \tag{9.1}$$

$$P(T|c_i) = P(a_1, a_2, \ldots, a_n, c_i) = P(c_i) \cdot \prod_{j=1}^{n} P(a_j|c_i) \tag{9.2}$$

The pros and cons of major machine learning methods are summarized in Table 9.4.

9 Local Energy Trading with EV Flexibility

Table 9.3 Economic parameters of generation units

Methods	Pros	Cons
Naïve Bayes	• Training only needs small datasets for training • Training is fast	• It assumes that all the features are independent • This algorithm is notorious
K-Nearest neighbours	• Versatility • Ease to use	• It starts to get extremely slow as the size of the dataset and the number of features increases
Support vector machine	• More effective in high dimensional spaces • Effective in cases where the number of dimensions is greater than the number of samples • relatively memory efficient	• Not suitable for large data sets • Does not perform very well when the data set has more noise
Decision tree	• Requires less effort for data preparation during pre-processing • Does not require normalization of data • Does not require scaling of data as well	• A small change in the data can cause a large change in the structure of the decision tree causing instability • Calculation can go far more complex compared to other algorithms
Random forest	• It deals well with the issue of overfitting • Very high predictive accuracy • Usability for a multitude of different applications	• Not easily interpretable • can be Computationally intensive for large datasets • Have very little control over what the model does
Deep neural networks	• Flexibility to be used for many applications • High prediction accuracy	• It needs large datasets to be able to perform well • It is very prone to over fitting
Recurrent neural networks	• Can model a collection of records (i.e. time collection) so that each pattern can be assumed to be dependent on previous ones • Even used with convolutional layers to extend the powerful pixel neighbourhood	• Gradient exploding and vanishing problems • Training is a completely tough task • It cannot system very lengthy sequences

9.5 Demonstration

Considering the impact of high renewable penetration in the wholesale market and the balancing market, this section investigates and demonstrates the various benefits by exploiting the EV flexibility in the dual market with different RER shares.

Table 9.4 Economic parameters of generation units

Generator	Efficiency-adjusted fuel cost (£/MW)	Non-fuel variable cost (£/MW)	Use of System charge (£/MW)	Carbon price (£/MW)	Short-run marginal price (£)
Gas-fired G1	2.22	3.38	1.53	16	23.13
CCGT G2	2.88	0.08	1.53	8	12.49
Gas-fired G3	15.38	1	1.53	15	32.9
Onshore wind G5	0	0	1.53	0.02	1.55
Offshore wind G6	0	0	1.53	0.02	1.55
PV G7	0	0	1.53	0.1	1.63
Biomass G8	0	1.4	1.53	1.8	4.0.74

The network topology of the used electricity system is shown in Fig. 9.7. This figure shows that the system includes four thermal units, i.e., gas-fired unit G1, gas-fired unit G2, and combined cycle gas turbine (CCGT). There are four renewable generation units, i.e., photovoltaic (PV), offshore and onshore wind turbines (WT), and biomass generation. Their economic parameters are as shown in Table 9.3. Their bidding prices in the day-ahead electricity market are defined by the efficiency-adjusted fuel cost, the non-fuel variable cost, the use of system charge, and the carbon price. The carbon price is calculated with the carbon density of different generators and the uniform carbon price so that the resulting price can align with the environmental target. The bidding prices in the balancing market are derived from the bid multipliers based on the history data.

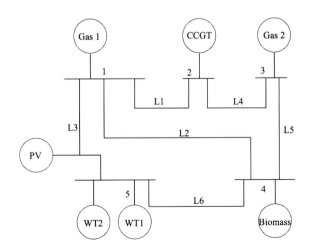

Fig. 9.7 Structure of the electricity system

9 Local Energy Trading with EV Flexibility

Considering the simplicity and feasibility of the proposed market model, the assumptions are made as follows:

- The day-ahead (DA) market is cleared through the merit order, considering the defined short-run marginal prices for different generators.
- The objective function of the balancing market is to minimize the total balancing cost due to load forecast deviation and renewable penetration.
- Five aggregators of EVs participate in the balancing market with different available volumes and bidding prices.
- In the balancing market, the bidding prices of all schedulable generators are assumed positive, which means that both upward and downward regulation causes balancing costs to the system operators rather than rewards.
- The clearing prices are calculated during each settlement period, i.e., 30 min.
- For simplicity, WT and PV are simulated as negative demands in the balancing market without considering their uncertainty.

9.5.1 Impact of RERs on the Electricity Market

To show the impact of PV and WT participating in the market, the day ahead prices are firstly simulated with various penetration percentages, as shown in Fig. 9.8. The penetration rates denote the proportion of PV and WT bidding in the DA market among the whole market (i.e., both the day-ahead and the balancing market). It can be found in the figure that participation of PV and WTs in the DA market remarkably reduces the system marginal price. With the penetration rate increasing, the periods of the peak price are shortened, while the periods of the valley price are extended. Therefore, the average DA price can be significantly reduced due to the low bidding price (i.e., short-run marginal price) of PV and WT in the market.

The average DA price and balancing cost are illustrated in Fig. 9.9. Aligned with the result in Fig. 9.8, the average DA price decreases with the growing penetration of PV and WT in the DA market, from around 28 £/MW·0.5 h. In contrast, the balancing cost increase with the increasing penetration rates of PV and WT in the balancing market. This is because the penetration of PV and WT is highly probable to increase real-time net imbalance volumes. The decrease rate of the DA price is lower and more volatile than that of the total balancing cost.

9.5.2 EV Flexibility Trading in the Balancing Market

In this part, the EV flexibility is traded in the balancing market by aggregators with different bidding prices, ranging from 0 to 30 £/MW·h. As shown in Fig. 9.10, incorporating EV flexibility aggregators in the balancing market can greatly decrease the average balancing cost. However, the performance varies with different bidding prices. Free EV flexibility leads to the largest cost reduction with around

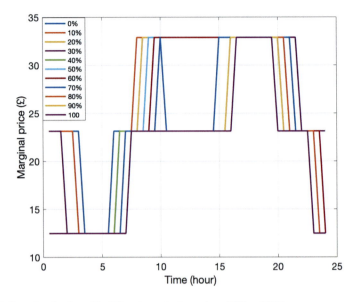

Fig. 9.8 Day-ahead price with different penetration rates of PV and WT

Fig. 9.9 Average DA price and total balancing cost with different penetration rates of PV and WT

16 £/MW·0.5 h for no PV and WT penetration in the balancing market and 16 £/MW·0.5 h for 100% penetration. In contrast, a negligible reduction of the balancing cost can be seen with the price of 150 £/MWh. The balancing cost reduction is negatively correlated with the EV flexibility price. When the penetration rate is 0%, only those prices lower than 30 £/MWh can notably reduce the average balancing cost. In contrast, when the penetration rate is 100%, the balancing cost can be reduced for all scenarios. That means the EV flexibility shows a more obvious contribution to reducing the balancing cost with more PV and WT participating in the balancing market.

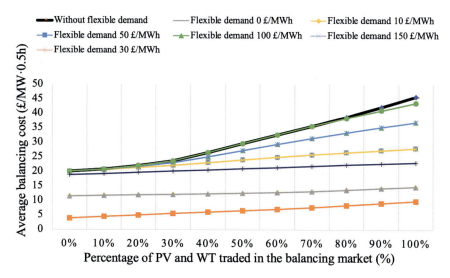

Fig. 9.10 Average balancing cost with various EV flexibility prices and PV/WT penetration rates

9.6 Further Discussions

In addition to the aforementioned contents, further investigation and research are required to facilitate flexibility markets but are not covered in this paper. Major future research interests are discussed as follows.

9.6.1 Carbon in Flexibility Markets

To materialize Net-Zero carbon ambitions, it is essential to create a flexible energy system to integrate a high volume of low-carbon technologies concerning power, heat, and transport technologies. In this paper, the carbon goals are incorporated in the local electricity market qualitatively. Nevertheless, specific quantification models of carbon emissions should be formulated in the flexibility market to demonstrate how much flexibility will be needed in a net-zero system. The potential research emphasis is shown below:

- Carbon signals: It is of great significance to identify the way of accounting for carbon emissions in flexibility markets through reasonable economic signals. Alternatively, carbon intensity should be measured and considered in all markets. Carbon pricing approaches, market-wide emission limits, and carbon evaluation in flexibility service procurement can also be incorporated in the market.
- Flexibility market decarbonization: It should be clarified the roles of energy system operators and distribution network operators in decarbonizing flexibility markets. For instance, additional considerations are needed in pricing frameworks,

e.g., RIIO2 (Revenue = Incentives + Innovation + Outputs) to favor decarbonized flexibility.
- Emission policies: Feasible energy policies should be deployed to help carbon emission reduction, e.g., carbon price support, emissions performance standard, and emission limits in the capacity market. Furthermore, the impacts of emission policies on flexibility should also be investigated for a stable energy system.
- Carbon monitoring: It is important to monitor the carbon intensity of flexibility markets, e.g., improve the transparency around carbon intensity of all energy markets. To materialize carbon monitoring, potential approaches are required, e.g., carbon intensity forecasting and ex-post carbon intensity reporting.

9.6.2 Flexibility Facilitation in the Multi-Energy System (MES)

Only taking the flexibility services into account in the electricity system restricts the development of decarbonization since the multi-energy system can provide considerable potential for demand-side flexibility, e.g., multi-energy carrier shifting and flexible demand for heating. Thus, it plays an indispensable part in decarbonization and the Net Zero economy. Energy system integration technologies should be utilized to explore the multi-energy system flexibility [42–44, 5]. For instance, energy coupling devices enable multi-energy load shifting in MES to provide operational flexibility. To unlock the flexibility in MES, corresponding management approaches are required, which are more complicated than those in the electricity system. In terms of storage technologies, hydrogen storage in the gas system and heat storage in the heating system also shows a good application prospect. Moreover, energy conversion resources can also be developed, e.g., power-to-gas, power-to-heat, and vehicle-to-grid.

9.6.3 Balancing Market Design

According to the dataset from BEIS (Department for Business, Energy & Industrial Strategy), with electricity demand reduction of around 15% in the early lockdown, the renewable penetration was very high, while the traditional flexible generation failed a lot. Therefore, the balancing action volumes remarkably increased for system reliability and stability requirements. The growing balancing actions turn down low carbon generation (e.g., wind and PV) and turn up traditional thermal generation (e.g., gas generation) which can provide operational flexibility for the system. Therefore, the carbon emissions increase with high renewable penetration, which runs counter to the environmental goals. Accordingly, ongoing market reforms are required for further carbon emission reduction, e.g., changes in the wholesale generation mix.

9.7 Conclusion

This chapter performed a thorough literature review of state-of-the-art research works in terms of EVs, DSF, and local electricity markets. By introducing the EV flexibility classification, its benefits to the electricity system are clarified. Afterward, the local flexibility markets are demonstrated based on their structure, including the centralized market, the decentralized market, and the P2P market. In addition, in line with appropriate flexibility evaluation models and market structures, potential trading and clearing technologies are introduced for local EV flexibility trading, i.e., blockchain, game theory, and machine learning. To illustrate the economic values of EV flexibility, a 5-bus local electricity system is used. Results show that renewable penetration in the wholesale market can significantly decrease the day-ahead price, while the growing penetration in the balancing market increases the balancing cost. Through integrating EV flexibility trading in the balancing market, the average balancing cost is notably decreased. Nevertheless, the reduction rates vary from the flexibility price of aggregated EVs. Finally, future research interests are discussed, i.e., carbon in flexibility markets, multi-energy systems, and balancing market design.

References

1. X. Yan, C. Gu, F. Li, Z. Wang, LMP-based pricing for energy storage in local market to facilitate PV penetration. IEEE Trans. Power Syst. **33**(3), 3373–3382 (2017)
2. A. Nikoobakht, J. Aghaei, M. Shafie-Khah, J.P.S. Catalão, Assessing increased flexibility of energy storage and demand response to accommodate a high penetration of renewable energy sources. IEEE Trans. Sustain. Energy **10**(2), 659–669 (2019)
3. X. Yan, C. Gu, F. Li, Z. Wang, LMP-based pricing for energy storage in local market to facilitate PV penetration. IEEE Trans. Power Syst. **33**(3), 3373–3382 (2018)
4. X. Yan, C. Gu, F. Li, Y. Xiang, Network pricing for customer-operated energy storage in distribution networks. Appl. Energy **212**, 283–292 (2018)
5. P. Zhao, C. Gu, Z. Hu, D. Xie, I. Hernando-Gil, Y. Shen, Distributionally robust hydrogen optimization with ensured security and multi-energy couplings. IEEE Trans. Power Syst. **36**(1), 504–513 (2021)
6. Z. Zhang, R. Li, F. Li, A novel peer-to-peer local electricity market for joint trading of energy and uncertainty. IEEE Trans. Smart Grid **11**(2), 1205–1215 (2020)
7. H. Shi, Q. Ma, N. Smith, F. Li, Data-driven uncertainty quantification and characterization for household energy demand across multiple time-scales. IEEE Trans. Smart Grid **10**(3), 3092–3102 (2019)
8. H. Shi, M. Xu, R. Li, Deep learning for household load forecasting—a novel pooling deep RNN. IEEE Trans. Smart Grid **9**(5), 5271–5280 (2018)
9. W. Golden, P. Powell, Towards a definition of flexibility: in search of the Holy Grail? Omega **28**(4), 373–384 (2000)
10. J. Ma, V. Silva, R. Belhomme, D.S. Kirschen, L.F. Ochoa, Evaluating and planning flexibility in sustainable power systems. IEEE Trans. Sustain. Energy **4**(1), 200–209 (2013)
11. E. Lannoye, D. Flynn, M.O. Malley, Evaluation of power system flexibility. IEEE Trans. Power Syst. **27**(2), 922–931 (2012)
12. I.A. Sajjad, G. Chicco, R. Napoli, Definitions of demand flexibility for aggregate residential loads. IEEE Trans. Smart Grid **7**(6), 2633–2643 (2016)

13. G. O. I. National Smart Grid Mission Ministry of Power, Electric vehicles in India. https://vik aspedia.in/energy/energy-efficiency/electric-vehicles-in-india
14. Y. Xue, X. Zhang, C. Yang, AC filterless flexible LCC HVDC with reduced voltage rating of controllable capacitors. IEEE Trans. Power Syst. **33**(5), 5507–5518 (2018)
15. G. Electric, What are the different methods of charging an electric vehicle? https://www.kia.com/eu/about-kia/experience-kia/technology/electrification/charging-methods-for-electric-cars/
16. D. Xie, H. Chu, C. Gu, F. Li, Y. Zhang, A novel dispatching control strategy for EVs intelligent integrated stations. IEEE Trans. Smart Grid **8**(2), 802–811 (2017)
17. S. Gottwalt, J. Gärttner, H. Schmeck, C. Weinhardt, Modeling and valuation of residential demand flexibility for renewable energy integration. IEEE Trans. Smart Grid **8**(6), 2565–2574 (2017)
18. P.J. Ramírez, D. Papadaskalopoulos, G. Strbac, Co-optimization of generation expansion planning and electric vehicles flexibility. IEEE Trans. Smart Grid **7**(3), 1609–1619 (2016)
19. A. Esmat, J. Usaola, M. A. N. Moreno, A decentralized local flexibility market considering the uncertainty of demand. Energies **11**(8), 2078 (2018)
20. A. Roos, Designing a joint market for procurement of transmission and distribution system services from demand flexibility. Renew. Energy Focus **21**, 16–24 (2017)
21. A. Bouorakima, J. Boubert, Y. Desgrange, Flexibilities in grid planning: case studies on the French distribution system. CIRED–Open Access Proc. J. **2017**(1), 2283–2286 (2017)
22. C. Li, Z.Y. Dong, G. Chen, J. Liu, F. Luo, Flexible transmission expansion planning associated with large-scale wind farms integration considering demand response. IET Generat. Trans. Distribut. **9**, 2276–2283 (2015)
23. T. Morstyn, A. Teytelboym, M.D. McCulloch, Designing decentralized markets for distribution system flexibility. IEEE Trans. Power Syst. **34**(3), 2128–2139 (2019)
24. H. Khajeh, H. Firoozi, H. Laaksonen, M. Shafie-khah, Peer-to-peer flexibility trading of end-users at distribution networks. pp. 797–799
25. C. Long, Y. Zhou, J. Wu, A game theoretic approach for peer to peer energy trading. Energy Proc **159**, 454–459 (2019)
26. W. Tushar, B. Chai, C. Yuen, S. Huang, D.B. Smith, H.V. Poor, Z. Yang, Energy storage sharing in smart grid: a modified auction-based approach. IEEE Trans. Smart Grid **7**(3), 1462–1475 (2016)
27. P. Shamsi, H. Xie, A. Longe, J.-Y. Joo, Economic dispatch for an agent-based community microgrid. IEEE Trans. Smart Grid **7**, 1–8 (2015)
28. Y. Zhou, J. Wu, C. Long, Evaluation of peer-to-peer energy sharing mechanisms based on a multiagent simulation framework. Appl. Energy **222**, 993–1022 (2018)
29. M.H. Cintuglu, H. Martin, O.A. Mohammed, Real-time implementation of multiagent-based game theory reverse auction model for microgrid market operation. IEEE Trans. Smart Grid **6**(2), 1064–1072 (2015)
30. Y. Rizk, M. Awad, E. Tunstel, Decision making in multi-agent systems: a survey. IEEE Trans. Cogn. Dev. Syst. **10**, 514–529 (2018)
31. W. Tushar, C. Yuen, D.B. Smith, H.V. Poor, Price discrimination for energy trading in smart grid: a game theoretic approach. IEEE Trans. Smart Grid **8**(4), 1790–1801 (2017)
32. S. Cui, Y.-W. Wang, N. Liu, A distributed game-based pricing strategy for energy sharing in microgrid with PV prosumers. IET Renew. Power Generat. **12**, 380–388 (2017)
33. C. Lo Prete, B.F. Hobbs, A cooperative game theoretic analysis of incentives for microgrids in regulated electricity markets. Appl. Energy **169**, 524–541 (2016)
34. D. Said, A decentralized electricity trading framework (DETF) for connected EVs: a blockchain and machine learning for profit margin optimization. IEEE Trans. Industr. Inf. **17**(10), 6594–6602 (2021)
35. N. Zhang, Y. Wang, C. Kang, J. Cheng, D. He, *Blockchain technique in the energy internet: preliminary research framework and typical applications*, vol. 36. (2016), pp. 4011–4022
36. Z. Li, J. Kang, R. Yu, D. Ye, Q. Deng, Y. Zhang, Consortium blockchain for secure energy trading in industrial internet of things. IEEE Trans. Industr. Inf. **14**(8), 3690–3700 (2018)

37. J. Kang, R. Yu, X. Huang, S. Maharjan, Y. Zhang, E. Hossain, Enabling localized peer-to-peer electricity trading among plug-in hybrid electric vehicles using consortium blockchains. IEEE Trans. Industr. Inf. **13**(6), 3154–3164 (2017)
38. S. Phelps, S. Parsons, P. McBurney, An evolutionary game-theoretic comparison of two double-auction market designs. pp. 101–114
39. G.E. Rahi, S.R. Etesami, W. Saad, N.B. Mandayam, H.V. Poor, Managing price uncertainty in prosumer-centric energy trading: a prospect-theoretic stackelberg game approach. IEEE Trans. Smart Grid **10**(1), 702–713 (2019)
40. R. Shipman, S. Naylor, J. Pinchin, R. Gough, M. Gillott, Learning capacity: predicting user decisions for vehicle-to-grid services. Energy Inform. **2**(1), 37 (2019)
41. M. Shibl, L. Ismail, A. Massoud, Machine learning-based management of electric vehicles charging: towards highly-dispersed fast chargers. Energies **13**, 5429 (2020)
42. P. Zhao, C. Gu, Z. Cao, Y. Shen, F. Teng, X. Chen, C. Wu, D. Huo, X. Xu, S. Li, Data-driven multi-energy investment and management under earthquakes. IEEE Trans. Industr. Inf. **17**(10), 6939–6950 (2021)
43. P. Zhao, C. Gu, Z. Cao, Z. Hu, X. Zhang, X. Chen, I. Hernando-Gil, Y. Ding, Economic-effective multi-energy management considering voltage regulation networked with energy hubs. IEEE Trans. Power Syst. **36**(3), 2503–2515 (2021)
44. Y. Shen, C. Gu, X. Yang, P. Zhao, Impact analysis of seismic events on integrated electricity and natural gas systems. IEEE Trans. Power Deliv. **36**(4), 1923–1931 (2021)

Chapter 10
A Review of the Trends in Smart Charging, Vehicle-to-Grid

Ridoy Das, Yue Cao, and Yue Wang

Abstract Electric vehicles have gained mainstream popularity and are hailed as a key technology that can abate the carbon footprint of transportation systems globally. However, there are fundamental challenges that can hinder widespread adoption, e.g. potential stress caused to the electricity network due to bulk-uncontrolled charging, battery degradation and ambiguous economic case for energy services. While Smart Charging and Vehicle-to-Grid are seen as potential solutions to these key issues, their suitability, especially concerning the latter technology is still today obscure, as a plethora of criteria influence its viability. This chapter identifies the most influencing parameters that determine the feasibility of Smart Charging and Vehicle-to-Grid, finds commonalities, and traces a trend through the past years with regards to the profitability of these two technologies. The research in this chapter shows that profits from Vehicle-to-Grid services can range from 13 to 207 £/vehicle/year, depending on the technical and economic factors, which, as demonstrated in this chapter, can vary considerably from one country to another.

Abbreviations

DSO   Distribution System Operator
EV   Electric Vehicle
ICE   Internal Combustion engine

R. Das (✉)
UK Power Networks, 237 Southwark Bridge Road, SE1 6NP London, United Kingdom
e-mail: ridoy.das@ukpowernetworks.co.uk

Y. Cao
Wuhan University, Wuhan 430072, China
e-mail: yue.cao@whu.edu.cn

Y. Wang
Scottish and Southern Electricity Networks, 1 Forbury Place, Reading RG1 2DQ, Perth, United Kingdom
e-mail: yue.wang@sse.com

© The Author(s), under exclusive license to Springer Nature Singapore Pte Ltd. 2023
Y. Cao et al. (eds.), *Automated and Electric Vehicle: Design, Informatics and Sustainability*, Recent Advancements in Connected Autonomous Vehicle Technologies 3, https://doi.org/10.1007/978-981-19-5751-2_10

MOO	Multi-objective optimisation
PHEV	Plug-in-hybrid electric vehicle
PJM	Pennsylvania-Jersey-Maryland
PV	Photovoltaic
RES	Renewable energy source
SOC	State of charge
UK	United Kingdom
USA	United States of America
USD	US dollar

10.1 Introduction

In this chapter, a review of the published research works on energy and ancillary service provision with electric vehicles (EVs) is presented. This is to establish the state-of-the-art in terms of EV integration, as well as to survey the profitability of energy services provided with EVs. Several works have dealt with EV and renewable energy source (RES) integration by looking at technical, economic and environmental aspects. However, researchers have encountered a number of hurdles while trying to quantify the profitability of Vehicle-to-Grid (V2G), evidenced by the wide span of results, ranging from very promising figures to some that depict V2G as unprofitable. The reason for such variable outcomes is due to the fact that a considerable number of factors come into play to decide the profitability of V2G, among which the most important ones are:

- technical aspects–i.e. technology status and constraints and
- economic parameters–i.e. tariffs, costs and payments, policy implications, supporting regulation.

The previous works on V2G are reviewed, by highlighting both the strengths but also the shortcomings of these studies to identify major gaps in knowledge. Subsequently, the most advanced research works on optimal EV charging scheduling are analysed separating the research on single-objective optimisation from those related to multi-objective optimisation. A few useful definitions are provided hereby to set the context of this review.

Definition 1 Vehicle to Grid is defined as "a system in which there is capability of controllable-bidirectional electrical energy flow between a vehicle and the electrical grid" [1].

Definition 2 When the energy flow is established between the vehicle and different archetypes, i.e., single household or a building, this service is called Vehicle to everything (V2X) charging/discharging.

10 A Review of the Trends in Smart Charging, Vehicle-to-Grid

Definition 3 Arbitrage is the "… purchase of a commodity or derivative in one market and the sale of the same, or similar, commodity or derivative in another market in order to exploit price differentials" [2].

10.2 Survey Motivations and Structure

When dealing with any technology that seeks commercialisation, two fundamental aspects are always examined: technical feasibility and economic viability. These two features represent the core of any successful and sustainable product. In this survey, the technology in question is the EV, with a focus on advanced charging strategy, hence their technical and economic features must be inspected. As there is both societal and political drive behind the development of EVs, environmental aspects are also taken into account, however, it is undeniable that eco-friendly, but unreliable and unproductive technologies are short-lived. Hence, a techno-economic feasibility assessment of EV charging strategies is of pivotal importance and constitutes the motivation behind this review.

EV charging strategies can be classified in uncontrolled ("dumb" or "dump") charging, smart (or "controlled") charging and V2G. The feasibility and benefits of smart charging compared to uncontrolled charging represent common knowledge among researchers and practitioners and includes cost savings, grid relief [3, 4] and improvement of battery life. On the other hand, the benefits of V2G are still today subject of heated debate, as will be evident from the results surveyed in this review. In fact, neither the level of prospective benefits that V2G can bring nor the elements that influence such level have been clearly reported yet. Consequently, individuals, user associations, industry, academia and policy makers are doubtful of the utility of V2G, which constitutes the greatest barrier for its wide implementation. There is an evident gap between the results achieved by academic research and industrial pilots and the final verdict on V2G. Consequently, more a more in-depth analysis is required for V2G rather than for smart charging.

Evidence of the public perplexity on V2G was effectively raised in [5], where 611 German drivers, including conventional internal-combustion-engine (ICE) powered vehicle and EV drivers, were surveyed on their willingness to participate to V2G services. Although the survey was conducted in 2013, the findings were published in 2018, and the majority of concerns and viewpoints still stand today. The topics covered by the survey were awareness of different EV types, elements that can enhance or limit willingness to participate to V2G, awareness of V2G and concerns and incentives to participate in V2G. They analysed the impact of several aspects characterising V2G services on participation and these were, plug-in restriction, minimum required range, possibility of indicating beginning and end of trips, different levels of monthly payments or one-off payments. The responses showed that most drivers were unaware of V2G, with only 1% declaring to know about it and that willingness to use a bidirectional charger was significantly less than that of using a unidirectional or even uncontrolled charger. This underwhelming response was due to the main concerns on V2G related to the prospective shortening of battery life, travelling pattern not being compatible to V2G services and that there will be third-party access to the vehicle which cannot be controlled, in order of importance. Enablers of V2G were overwhelmingly dominated by cost related aspects, i.e., cheaper charging compared to uncontrolled charging, discounts on purchasing an EV or a charging station and an annual bonus. By applying ordinal regression, the authors found the impact of the combinations of these factors: the results indicated that drivers expected high payments (compared to conventional electricity tariffs) to reduce their minimum driving range requirements and allowing an on-board computer.

Addressing a different category of stakeholders, in [6], 227 experts were queried on the benefits of EVs and V2G. Participants from 200 institutions in Denmark, Finland, Iceland, Norway and Sweden, the likes of national and government ministries, universities and research institutions, electricity transmission and distribution utilities, car manufacturers, private companies and industry groups and associations, were interviewed. They gathered the opinions of important names in different fields, such as BMW, Volkswagen, Nissan, E.ON, Tesla Club and pioneers in the field of smart charging and V2G, such as Fortum and Nuvve. Unsurprisingly, the experts perceived the environmental benefits of EVs as major drivers: reduced emissions, followed by reduced noise, better performance and only then economic savings and more integration with renewables were mentioned. Mirroring the outcome of [5], the overall knowledge on V2G was more limited, with only 66% of the experts discussing the benefits of V2G. The majority of the experts identified the possibility of integrating with intermittent renewable energy as a key benefit. Moreover, V2G was comparatively more often linked with domestic solar than wind, with experts saying it is a more intuitive connection. Smart (controlled) charging was seen as the second most popular advantage being also defined as a steppingstone for V2G. Those that were aware of the economic benefits of V2G, agreed on similar levels of earning of around 120 euro/month (107 £/month).

Comparing the findings of the two studies, they surveyed the two sides of the debate, users and specialists. One common aspect is the relatively limited awareness of the V2G concept; even though the material from [5] are based on the situation in 2013, the currently limited number of V2G implementations indicates that awareness did not improve much from then. Understandably, users were mainly concerned about factors that directly relate to them, such as travelling patterns, battery life and cost reduction. Experts were more informed about wider objectives, such as reducing intermittence of RES. In addition, a rather good estimate of potential profits was brought forward. From this brief, yet illuminating scrutiny, two research questions are raised:

- Is V2G currently profitable?
- What are the factors that influence the profitability of V2G and what is their impact?

10.3 Survey on Economic Feasibility of V2G

With the aim of responding to the research questions defined in Sect. 2.3.1 papers have been collected and reviewed. These have been retrieved from the Google Scholar search engine as it collects research papers from the major publishers including IEEE, Elsevier, Nature, Francis and Taylor, Wiley, MDPI among others. The collection research works spans over a period of 13 years, from 2007 to 2019, to provide a chronological roundup of the advancements in this field. A few rules are established for a coherent and rational investigation, for this review and throughout the thesis:

– Some pioneering research works are referred to regardless of the year of publication; this is because such works were the first in initiating the research in that area and they serve as references for the most updated research.
– As indicated in the introduction of this chapter, this survey will deal with technical and economic aspects in the area of EV charging strategies. While cost factors are heavily influenced by the time of publication, as the economic parameters, policies and market status can change significantly in a matter of few years, technical performance is a does not change significantly in a matter of few years. For instance, if EVs are optimally scheduled to reduce peak electricity demand by a certain amount, the magnitude of this reduction will not change across some decades. On the other hand, economic benefits change as the influencing factors vary in time. We therefore provide a chronological roundup of the works that dealt with economic aspects related to smart charging and V2G, while for technical achievements, i.e. peak shaving, voltage balancing, the time dimension is not a concern.
– Cost values were all converted to British pounds to allow comparative analysis.

Literature [5–21], provided some insights on the economic dimension of V2G, while literature [22–36] dealt with technical aspects. Table 2.3.1 summarises the settings considered in [5–21] in chronological order. The factors that are highlighted have been categorised based on criteria set hereby:

- Technological and market considerations

  – Time; as technology advances ad reaches mass production, cost comes down, markets saturates, all leading to different implications on prospective benefits through years.
  – Country, market and service; different countries will have different policies in place and different market structures designed for the various V2G services.

- Case-study setup

  – EV battery capacity; EVs of different categories, with diverse battery capacities can be utilised to provide V2G services, and since service payments are often proportional to the energy exchanged, this factor is crucial in determining potential remunerations.

- Charger rating; several services, including frequency regulation provide payments that are proportional to the committed power and EV chargers, in combination with on-board power electronics, decide the feasible power level.

- Cost–benefit considerations

  - Battery investment cost: this is one of the most critical factors in determining the prospective benefits. In fact, as increased utilisation from V2G is known to lead to battery wear, the underlying battery cost discerns the economically feasible services.
  - Charger cost; the cost of a V2G charger is a cryptic information and it is a fixed cost that can weigh on the cost–benefit calculation.
  - Electricity tariff and service payment; depending on the type of service, V2G can be employed to reduce electricity bills or provide ancillary services. In the former case, the (avoided) electricity tariff constitute the main revenue stream while in the latter, it is the service payment.

- Considerations on a realistic assessment

  - Battery degradation model; model simulations provide estimates of the real-life operation. By using battery degradation models, a more accurate account of the real cost-benefits can be given.
  - EV availability: as EVs are primarily used for transportation, their unavailability as parked and plugged-in assets will definitely affect the achievable profits

In Table 10.1, shaded boxes indicate that the associated information was not provided/considered in the study. It should be noted that many of the considered features coincide with the main points indicated by [5] (battery cost and degradation, EV availability, cost consideration etc.) and [6] It can be seen from the same table that the chronological roundup starts from [7], where the foundations of V2G implementation for ancillary services were first laid. The V2G concept was first academically introduced in 2005 by Professor Willet Kempton based at the University of Delaware, USA. His team defined the basic setting for the economic viability assessment for EV fleets providing network services. They simulated frequency regulation provision in the Pennsylvania-Jersey-Maryland (PJM) market for a fleet of 250 vehicles, and calculated revenues in the range of USD 427–3,555 per vehicle. It was argued that the wide spectrum of profits is determined by three factors: the rating of the charger, the energy stored in the battery (if the battery of an EV is either empty or full, then it cannot provide the entire regulation up and down service) and the number of available EVs. The upper bound of their calculated profits is comparatively high, when compared with more recent studies, as can be seen from Table 10.1 and Fig. 10.1, where whenever required, profit ranges have been used to report the results.

Only [13] provided a profit higher than [7], which however is due to the high battery capacity of the considered buses (80–108 kWh) and committed power (chargers rated 70 kW). Even [8], which assessed the economic feasibility of frequency regulation provision in the same market by assuming 24 h EV availability,

Table 10.1 Summary of research works on feasibility of V2G

Literature	Profits (£/vehicle/year)	Service	Battery capacity (kWh)	Charger rating (kW)	Battery cost (£/kWh)	Charger cost (£)	Electricity tariff (ET) or regulation prices (RP) (£/kWh)	Battery degradation consideration	EV availability consideration
Tomić and Kempton [7] (2007) USA	427–3,555 depending on the committed power	Frequency regulation	11.5, 27.4	2.9, 6.6, 15	277, 479	435	ET = 0.04 RP = 0.006–0.03	Degradation model function of DOD	• Commuter fleet with availability during the day and night • Utility fleet available 3 pm to 8 am
Han et al. [8] (2012) USA	[910 1,529 2090]	Frequency regulation	4.25, 14.5, 40	Char. 60 Dich.120	PHEV 750 BEV 237–395	**Not considered**	RP = 0.03	Degradation according to testing cycles set by OEM	**Full availability**
De Los Rios et al. [9] (2012) USA	988–1106	Frequency regulation	20, 99	19.2	790–948	1,580–3,160	**Not stated, New England regulation market**	Degradation model function of DOD	Availability from 8 pm to 8 am
Schuller et al. [10] (2014) Germany	108–207	Energy arbitrage	28.3, 35, 53	3.3, 11	119–483	131,476	ET = 0.18 Wholesale price = 0.0612	Degradation model considering power and energy fading	Availability of employed users and retired ones

(continued)

Table 10.1 (continued)

Literature	Profits (£/vehicle/year)	Service	Battery capacity (kWh)	Charger rating (kW)	Battery cost (£/kWh)	Charger cost (£)	Electricity tariff (ET) or regulation prices (RP) (£/kWh)	Battery degradation consideration	EV availability consideration
Agarwal et al. [11] (2014) Singapore	Frequency regulation: 600 Primary reserve: 0 Secondary reserve: 21 Contingency reserve: 370	Frequency regulation, primary, secondary and contingency reserve	24	6.6	214	**Not considered**	RP = 0.054, 0.00026, 0.0011, 0.0091	Degradation model function of DOD	Different availability depending on the employment status
Zeng et al. [12] (2015) USA	347–1,251	Frequency regulation	7.6, 23, 60	3.3, 6.6, 10	237	830	0.0079–0.0711 + 0.004	Degradation model function of DOD	15 h of provision per day (365 days)
Shirazi et al. [13] (2015) USA	6,909, 6,997	Frequency regulation	80, 108	70	**Not considered**	23,700	ET = 0.079–0.082 RP = 0.0237 (std. dev. 0.0103)	**Not considered**	180 days 5–8 am and 2–5 pm, and 185 days of full 24 h provision
Ciechanowicz et al. [14] (2015) Singapore	Frequency regulation: 40.3–51.4 Primary reserve: 0 Secondary reserve: 0 Contingency reserve: 13	Frequency regulation, primary, secondary and contingency reserve	20	40	436	**Not considered**	ET = 0.158 (std. dev. 0.0032) RP = 0.0521 (std. dev. 0.0229), 0.0002 (std. dev. 0.0013), 0.0008 (std. dev. 0.0025) and 0.0065 (std. dev. 0.0036)	Comprehensive battery degradation model function of DOD and SOC	Travelling patterns of commuters in Singapore were considered

(continued)

Table 10.1 (continued)

Literature	Profits (£/vehicle/year)	Service	Battery capacity (kWh)	Charger rating (kW)	Battery cost (£/kWh)	Charger cost (£)	Electricity tariff (ET) or regulation prices (RP) (£/kWh)	Battery degradation consideration	EV availability consideration
Zhao et al. [15] (2016) USA	1,563, 2,487	Frequency regulation	40, 80	19.2, 25	474	5,451–5,609	ET = 0.0632–0.11 RP = 0.013–0.039, 0.0073–0.0239, 0.0093–0.047, 0.0087–0.0301, 0.0084–0.0324, depending on the region	A total battery life cycle for V2G of 2000–6000 cycles is considered	Assumption of 60% of the total available range left for V2G
Peng et al. [16] (2017) USA	1298	Frequency regulation	16, 24, 35	8, 10, 12	**Not considered**	**Not considered**	ET = 0.0008–0.0261 RP = 0.004–0.0221	**Not considered**	Travelling patterns generated from statistical data
Kuang et al. [17] (2017) USA	73–2,442, depending on the demand profile and EV initial charge	Load levelling	30	1.5, 7.5	**Not considered**	**Not included**	**Time of Use, not specified**	**Not included**	Daytime availability with long and short duration and night time availability

(continued)

Table 10.1 (continued)

Literature	Profits (£/vehicle/year)	Service	Battery capacity (kWh)	Charger rating (kW)	Battery cost (£/kWh)	Charger cost (£)	Electricity tariff (ET) or regulation prices (RP) (£/kWh)	Battery degradation consideration	EV availability consideration
Gough et al. [18] (2017) UK	167–1,367	Short time operation reserve, energy arbitrage, capacity market and triade avoidance	24	12	160	3,750	ET = 0.085; STOR availability payment = 0.0033 – 0.0043; STOR utilisation payment = 0.167–0.171	A total battery life cycle for V2G of 1020 cycles is considered	Fleet of utility vehicles with availability from 7 am to 11 pm
Tamura and Kikuchi [19] (2018) Japan	Not profitable. Requires breakeven price of 0.44–1.83 £/kWh	Frequency regulation	40	Not specified	283	Not considered	System price = 0.071	Degradation model with SOC dependency	**Full availability**
Liu and Zhong [20] (2019) China	Not profitable, levelised cost of storage is higher than commercial electricity tariff	PV integration	60	7	130	224	ET = 0.09	Model based on utilisation	Availability during the day at working hours
Datta et al. [21] (2019) Australia	135	Electricity cost reduction	23	Not specified	**Not specified**	Not considered	Peak ET = 0.19 Off-peak ET = 0.095	**Not considered**	Residential pattern: EV unavailable during the day on weekends

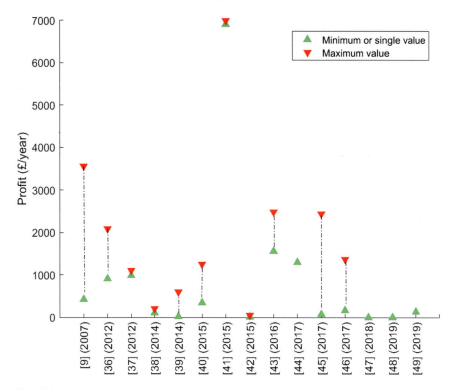

Fig. 10.1 Profits from V2G provision in chronological order

reported lower profits than [7]. It can be seen from Table 10.1 that both works used similar capacity payments, but the latter showed much lower profits, despite the very optimistic availability assumption. This is due to the more realistic assumption on battery cost, which was the second highest in [8]; high battery cost weighed heavily on the achievable profits. Interestingly, the highest battery cost was adopted in [9], which also assessed frequency regulation provision and was conducted in the same year as [8], indicating that these values of battery cost represented the most sensible levels at the time. Considering that [7] was published five years earlier than [8, 9], it can be concluded that the former assumed a rather unrealistic value of battery cost (as well known, manufacturing costs decrease in time driven by increased scale of production). More recent works, for instance from 2016 onwards, exhibit a sharp decline in profits. Comparing the results achieved in [8, 9] with those from more recent studies, the closest one is [16], published four years later. However, they reported higher profits than the studies in 2012. This may be due to the fact that they did not consider battery degradation in their calculations; considering the same battery investment cost and total V2G cycles as [15], published one year earlier (therefore battery cost should not be much different), namely 474 £/kWh and 4000 cycles respectively, the cost of degradation comes at 0.11 £/kWh, which is more than threefold the service payment they considered, making the service not profitable. In fact, [15] resulted as a profitable business case because they employed both higher service payment, battery capacity and committed power. This may indicate that in some USA markets the payments can be less favourable to EV fleets providing frequency regulation now, as compared to 2012.

In addition, the striking popularity of frequency regulation is evident, with two thirds of the works investigating its profitability. Almost equivalent is the recurrence of the USA in the list of countries (eight times out of fifteen); in fact, all the works that dealt with markets based in the USA chose frequency regulation as prospective service. This highlights the effect that supporting regulatory and policy frameworks can have on the adoption of technologies. In fact, as stated in [12], the PJM market provided two types of signals: the conventional regulation signal, denoted as RegA, for conventional power plants and performance-based regulation signal, denoted as RegD, for assets with fast response capability. The latter provided a capacity payment, that is proportional to the committed time, and a performance payment, proportional to the ratio between the variability of the RegD signal and the variability of the RegA signal. This is particularly favourable towards batteries, which are inherently characterised by superior response capability. In contrast, as stated in [19], in 2018 a frequency regulation market was not yet available in Japan, which contributed to the nil profits stated in the same study. The only other country where frequency regulation has been considered is Singapore, with [11, 14], both reporting lower.

profits than works that focused on the USA market. This is due to the comparatively lower payment for ancillary services (especially in [14]).

Figure 10.2 represents the battery investment cost utilised in [5–21].

In the above figure, a general decreasing trend is seen for the costs; the battery cost starts at nearly 1000 £/kWh in the yearly 2010's, going down to under 400–300 £/kWh at the end of the decade. This easy comparison further evidences the irregularly low battery degradation cost utilised in [7].

Figure 10.3 shows the service payments employed in [5–21].

It may seem from the figure above that there is an increase in the service payments, however, this is mainly because the studies from 2017 either focused on other services (reserve, bill reduction, PV integration) or different countries (UK, Japan, Australia), or alternatively frequency regulation resulted not profitable. Due to the diversity of services, costs and model assumptions, [5–21] were clustered to find similarities in their analysis according to the procedure outlined below:

- A total of 11 features were identified for the 15 studies [5–21]. All features are sequentially combined to determine a characteristic signature for each work.
- Each feature is normalised to the maximum value achieved by the research works along that feature. Therefore, the maximum value that a feature can achieve is 1.
- Reference [13] has been removed due to its unnaturally high profits, which was due to the high energy and power committed. As it was an isolate case, it has been removed.
- Year of publication starts at 0 for 2007, [7] to 1 for 2019.

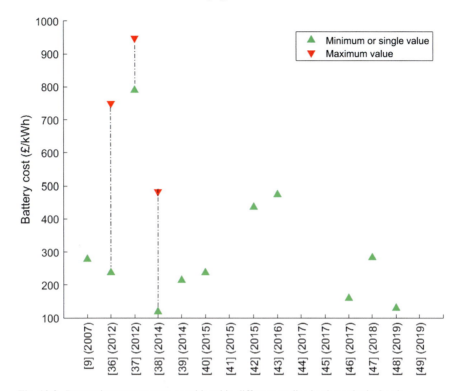

Fig. 10.2 Battery investment costs considered in different studies in chronological order

- Countries are identified as [0.143, 0.288, 0.429, 0.571, 0.714, 0.857, 1] representing {China, Singapore, Japan, UK, Germany, Australia, USA} respectively.
- Services are identified as [0.333, 0.666 1] representing {Energy arbitrage, Demand provision, Ancillary services} respectively.
- Battery degradation consideration has been categorised as [0 0.333 0.666 1] representing studies that did not consider battery degradation, studies that considered a fixed number of charging/discharging cycles, models that considered one impacting parameter and models that considered more than one impacting parameters, respectively.
- EV availability consideration has been categorised as [0 0.5 1] representing studies that considered EVs as always available for V2G services, studies that considered a fixed availability pattern and finally studies that considered real-life patterns based on data.

Six clusters were chosen as a right trade-off between diversity and number of studies per cluster. The results are presented in Fig. 10.4, where the lead author and the year of publication are reported for each study.

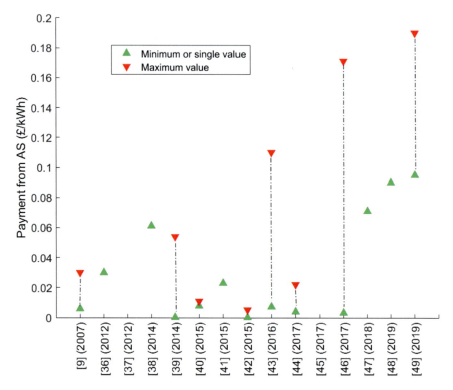

Fig. 10.3 Service payments considered in research works in chronological order

The following discussion is structured based on four points:

- *Effect of time*—as technologies mature and reach mass production level, the associated costs decrease, making those technologies more profitable. This has been the case for battery degradation: in fact, all the clusters, with the exception of cluster 4 (blue curves) contain studies from a similar period of time and by comparing clusters 1, 2, 3, 4 and 6 a decreasing trend of battery degradation is seen. In addition, with time, as certain services become popular, the associated markets tend to saturate, leading to lower payments. That has been the case for frequency regulation in the USA; in fact, [16] compared to [7] reported a lower upper bound for the capacity payment.
- *Influence of the market*—the different market options in place, along with their regulations are crucial in promoting or discouraging the adoption of a certain technology. Again, that has been the case for frequency regulation, which has been very popular in the USA and EVs were encouraged to participate. In fact, unsurprisingly most of the works that addressed frequency regulation dealt with the USA markets. However, as investigated in [15], various markets will provide different payments, and when the market is not made available, the service results unprofitable, as was the case for [19]. Another example is [18], which evaluated

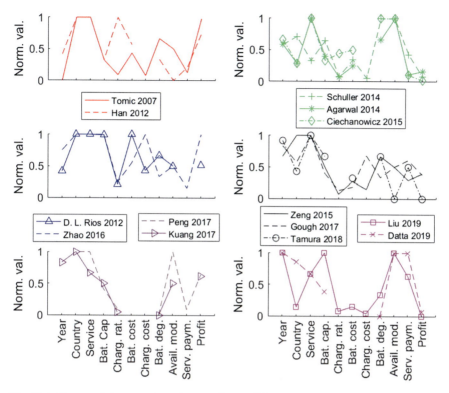

Fig. 10.4 Clusters of the research works according to different parameters

different ancillary services in the UK. Different services require different participation requirements: from few calls per year, i.e. reserve markets and capacity market, to several calls per day, i.e. frequency regulation. Participation requirements, along with the level of payment provided for power (i.e. capacity) and energy will determine the feasibility of a service. In fact, in [18], energy arbitrage was not enough to provide a successful business model; capacity market or triad avoidance, both providing relatively high capacity (power) payment were necessary to improve the benefits.

- *Impact of battery degradation*—this, along with service payment, is the main variable cost for V2G service provision. As already discussed, the cost of lithium-ion batteries has reduced over time, with current prices in the range of £/kWh 150–400. However, as demonstrated in current literature [10, 11, 14, 20] battery cycling inevitably leads to battery degradation. Any cost–benefit analysis aimed at assessing the feasibility of V2G services, needs to appropriately model and estimate prospective battery degradation incurring from service provision. To this end, some only [10, 14] modelled battery degradation with two or more impacting parameters, i.e. Depth of Discharge, State of Charge, charging rate, [9, 19] considered only one parameter, while the remaining works considered only the number of cycles as impacting factor or did not include battery degradation in their model. It can be seen that as accuracy of battery degradation increases, the corresponding profits decrease: compare for example [10, 14] with [11], all in the same cluster (number 2, green curves in Fig. 10.4). In fact, [11, 14] provide the same service in the same country (Singapore), but the latter reported much lower profits due to a more accurate battery degradation estimation. Based on the results achieved in the available literature and depending on the magnitude of the impacting parameters mentioned above, we estimate a battery degradation cost in the range of £/kWh 0.075–0.3.

- *Impact of EV availability model*—as storage operation is only a secondary function that EVs can serve, the impact of different transportation, consequently availability, models need to be considered in economic assessments of V2G services. However, much like battery degradation, it was at times neglected [8, 19], or fixed availability was assumed [7, 9, 12, 15, 17, 18]. Only [10, 14, 16, 20, 21] simulated the actual travelling pattern of EVs based on historical data. With the only exception of [16], which did not consider battery degradation, the remaining studies ([10, 14, 20, 21]) reported low profits from V2G. Realistically, when considering the impact of all the factors highlighted in this review, V2G can bring £/vehicle/year 13–207.

10.4 Survey of the Impact of Smart Charging and V2G on Distribution Networks

Smart charging and V2G can help to mitigate the impact of bulk and uncontrolled EV charging, and consequently help to accommodate a higher share of EVs in the national vehicle fleet interacting with the electricity grid. More generally, smart and V2G charging from EVs could help to achieve an efficient utilization of the grid by addressing peak demands, integrating more intermittent renewable energy power and filling in the load curve in hours characterized by low power consumption. This can potentially lead to grid investment deferral. Following this idea, [22] evaluated the potential benefits for prospective Distribution System Operators (DSO) from investments in V2G services and compared them with the underlying grid investments. The authors inferred that there is a certain potential of peak electricity demand reduction resulting from a number of EVs providing peak shaving service. This in turn affects the duration curve of the network which depends on the electricity demand profiles. Ultimately, a balance is struck between the number of operational hours of storage, which determines battery degradation cost, and the avoided network investments. With 250 EVs, they showed that there was the potential of reducing the peak demand by 900 kW, by using 3.6 kW chargers. By considering a degradation cost of £/kWh 0.18 (resulting from a battery investment cost of £/kWh 267–623), they showed that below an annual energy throughput of 135 MWh/year, the avoided grid investments achieved by V2G were higher than the incurred battery degradation. However, they argued that with an average spot electricity price of euro/kWh 0.027–0.062 provided in North European countries, the economics of V2G did not make sense, as energy could have been bought from the wholesale market in order to satisfy the peak. However, as discussed in Sect. 10.3, the cost of lithium-ion batteries is currently in the range of £/kWh 150–300, and the associated battery degradation cost is £/kWh 0.075–0.3. Hence, peak power provision can become a profitable service in the near future. It should be noted that [22] did not consider the travelling patterns of EVs, which would reduce the potential peak demand reduction with V2G, nor the cost of V2G chargers. Hence, an economic analysis of peak shaving, and the associated benefits that DSOs can reap, must be conducted. EVs can also be charged by imposing network constraints as was demonstrated in [23]. They tested the operation of a multi-agent system in a laboratory setup, where one EV was emulated by hardware in loop and 60 EVs were simulated. The emulated EV complied with network constraints.

References [24–28] further investigated the potential peak reduction capability of EV fleets equipped with V2G. In [24], the effect of smart charging on the electricity demand profile of a distribution network was analysed. The EVs were connected through a level 2 charging, either at home or in public areas, where renewable energy from PV and wind was available. 50,000 EVs performing smart charging enabled a peak demand reduction of 87 MW.

The location where information is stored, and hierarchy of computation can influence the potential achievable grid relief. In fact, measurements for an entire distribution network can be collected and utilised in a central server, or the decision-making privilege can be shared among multiple agents, distributed in the network. References [24, 28] evaluated the difference in these two strategies by exploiting intelligent EV charging to perform peak shaving and reduce the variability of the load profile in a local distribution grid. Local and global control strategies were performed and compared to a business-as-usual scenario with uncontrolled charging. Future scenarios with different PHEV penetration level were simulated and these are 15, 45 and 75%. Given the nominal voltage level of $230 \pm 10\%$, uncontrolled charging led to more voltage deviation. Scenarios simulating a 10%, 30% and 60% of PHEV penetration rate were considered. The local control strategy let to improvements in peak demand in the order of 8–38% compared to the business-as-usual case, while the global strategy achieved 8–42% of improvement. Both the local and global energy control strategies improved the flatness of the load profile, but the global energy control strategy resulted in the most optimal load profile. Although global control strategies provided the highest improvements in peak demand, it should be noted that the implementation cost of a centralised control strategy is disproportionately higher than that of a decentralised system, due to the onerous communication infrastructure. The additional grid relief given by a centralised architecture must be compared against the incurring costs when choosing between the two strategies. It was further evidenced by [24, 28] and confirmed by [27, 29] that the penetration rate of EVs brings an additional dimension when evaluating grid benefits. The authors of [27] evaluated the provision of peak load support as well as voltage unbalance mitigation in a cluster of three feeders of a distribution network in Australia. They showed that above a rate of 40% EVs being available for those network services, there are beneficial effects in terms of voltage rise mitigation. In [29], for 25 and 50% EV penetration levels in a distribution network (corresponding to 31,250 and 62,500 EVs) it was shown that uncontrolled charging increased the peak demand by 36 and 74%, respectively. However, the benefits also scaled up proportionally as smart charging achieved peak levels that were 13 and 27% lower than those caused by uncontrolled charging.

As reported in [22], the category of the electricity demand profiles will have a substantial influence on the potential peak demand reduction achievable by EVs. For instance, if the load duration curve of a network exhibits a substantially high peak compared to its base demand, then EVs have to provide V2G support for a limited number of hours per year and targeted to critical moments. Conversely, if the load duration curve is flatter, than the EVs must be available for longer periods in order to achieve some peak demand reduction. This aspect was investigated by [26] where three case studies, namely high-rise residential buildings, office buildings and commercial buildings were analysed to quantify the benefits of peak shaving. 15 EVs achieved a peak demand reduction of 9.34–10.62%, 27.21% and 15.25%, respectively.

Few works evaluated the benefits of V2G for behind-the-meter services [30, 31]. In [30], the possibility of integrating EV charging with the energy generated by PV systems and a backup solution in case of emergency conditions were analysed. They applied their energy management strategy to a commercial building with 220 office-working stations, a 341.6 kWp PV installation, a 60kWh stationary storage and 48 EVs. The results showed that V2G can optimally integrate with PV by charging during periods of excessive generation and supplying the evening demand. They also validated backup provision in emergency conditions. Similarly, in [31], a model for grid stabilisation with 250 EVs residential/commercial buildings in Brazil was developed. A three-level tariff was considered for the case of peak demand reduction, while the variability of the net power exchange was minimized to improve grid stability. However, they found that optimising grid stability does not lead to the maximum profit for users, which further emphasises the need for Multi-objective optimisation (MOO) strategies.

10.5 Transportation Compatibility for V2G Implementation

From the surveyed works, it is evident that the economic potential of V2G depends on several factors, which have been discussed individually in Sect. 10.3. Crucially, the underlying influence of both driving requirements and EV charging behaviour impose strict constraints, as in order to be available to provide V2G services the EV needs to be both parked and plugged in. Due to the usage of the EV for transportation, and the associated charging or battery state of charge requirements, it is not always possible to provide grid supporting services. As already emerged from analysing the literature in Sect. 10.3, accurate prediction of EV travelling patterns is pivotal in accurately estimate V2G profitability and develop profitable business models. In fact, [10, 18, 26] evaluated the availability of EVs for providing certain V2G services and found that some service provisions are limited by the characteristics of the users' driving pattern. To this end, the research works analysed thus far have addressed this requirement with different methods: by considering EVs parked for 24 h, by considering a fixed availability period or by simulating realistic travelling patterns from historical data.

Full availability. Studies such as [8, 19, 22], assumed that EVs were always available for V2G services. While this can provide the maximum achievable profits/benefits, it is not an accurate representation of the results achievable in real-life conditions.

Fixed availability. Several studies, [7, 9, 12, 13, 15, 17, 18, 27, 31] considered certain availability periods where all simulated EVs were made available for V2G services. Although this model is better than the previous strategy, in real-life operation, unless contracted, not all EVs will comply with set availability period. In fact, there is always diversity in EV patterns, where different plug-in/plug-out times and energy requirements must be simulated.

Random availability. As discussed in Sect. 10.3, [10, 14, 16, 20, 21] randomly generated diverse EV travelling patterns, with associated plugging-in/out times and charging requirements. In addition, [28] simulated EV availability based on random plug-in and out times normally distributed around 17:30 and 6:30 respectively with a standard deviation of 45 min. [25] adopted a forecasting model for the energy required by EVs based on US driving patterns. In particular, the work done in [26] was effective in modelling the impact of different driving patterns. For instance, EVs were not available at residential buildings during the office hours. For office buildings, the EV availability patterns were opposite to that of residential buildings. Differently from the previous two cases, in the case of commercial buildings, the travelling patterns can be considered known ahead as they depend on the tasks that need to be carried out, i.e., for postal delivery.

Five notable works accurately estimated EV availability patterns using probabilistic methods. In [32] the operation of different size of EV fleets was analysed in order to both satisfy the EV charging requirements and to provide frequency regulation. The distribution of the driving schedules was randomly sampled from real-life EV usage data and information on the daily driving routine. They estimated that the probability of having a high availability of EV battery capacity was high during the night, in the early morning and at late evening. These profiles were compared against the frequency regulation capacity requirements; as a result, the probability to meet a certain grid-facing bid requirement and to bid the optimal grantable capacity taken up and paid for by the grid were calculated. This represented an exemplary approach in considering EV availability for V2G purpose. Similarly, in [33] the potential EV power capacity to provide frequency regulation was estimated. Different factors including probability of EVs arriving at the parking spot at certain states of battery charge, in terms of initial SOC and required SOC, time of arrival and planned departure time, and a queuing system for different services were considered. EVs were assigned to the types of grid services depending on their availability and the capacity available for that service was estimated. In [34], a Markov Chain Monte Carlo method was employed to extract the trip and idling time information from real-life vehicle driving data. This study found that EVs are driving, parked at home, parked at workplace and parked at other places 5.2%, 59.6% 33.6% and 1.6% of the time respectively. In [35], due to the high probability of the EV being parked at home or workplace, the authors considered these as charging locations. The simulated synthetic driving pattern then fed into a V2G scheduling, aimed at household peak shaving. An equally effective estimation model based on a queuing system was employed by [36]. The charging requirement of the vehicles was modelled with a queuing system based on a random probability distribution for each vehicle. Then the stochastic net demand for the parking lot was calculated from such probabilities. Random availability models allow an accurate estimation of the influence of EV travelling constraints on V2G benefits; in fact, as was shown in Sect. 10.3, and emphasised by [14], whenever randomised travelling patterns were considered, the associated benefits from V2G were reduced from excessively optimistic figures to realistic levels.

10.6 Conclusions

This survey presented the recent trends on V2G technologies in terms of their profitability and the most advanced optimisation techniques to integrate EVs and RES by achieving different objectives. Based on the analysis of the results from the presented studies and real-life demonstration projects a number of key conclusions can be drawn:

- Smart charging and V2G services provide benefits both from an economic and a grid operation point of view. Additionally, Smart charging and V2G reduce the impact of EV charging on the optimal grid operation.
- Battery cost, service payment and availability of EVs are major influencing factors that decide the economic viability of V2G services.
- Battery degradation cost falls in the range of £/kWh 0.075–0.3 and profits from V2G services can range from 13 to 207 £/vehicle/year, depending on the technical and economic factors.
- To allow V2G profitability, EV batteries must become more economically competitive, and they should currently be exploited as power sources, i.e. high powers and low energies actually provided.
- Aggregators will play a primary role as intermediary between V2G service providers at different scales and stakeholders that demand such services.

References

1. A. Briones, J. Francfort, P. Heitmann, M. Schey, S. Schey, J. Smart, *Vehicle-toGrid power flow regulations and building codes review by the AVTA*, (Idaho National Laboratory, 2012)
2. https://www.ferc.gov/market-oversight/guide/glossary.asp
3. G. Putrus, E. Bentley, R. Binns, T. Jiang, D. Johnston, Smart grids: energising the future. Int. J. Environ. Stud. **70**, 691–701 (2013)
4. G. Fitzgerald, J. Mandel, J. Morris, H. Touati, *The Economics of Battery Energy Storage- How Multi-Use, Customer-Sited Batteries Deliver the Most Services and Value to Customers and the Grid*, (Rocky Mountain Institute, 2015)
5. J. Geske, D. Schumann, Willing to participate in vehicle-to-grid (V2G)? Why not! Energy Policy **120**, 392–401 (2018)
6. L. Noel, G. Zarazua de Rubens, J. Kester, B.K. Sovacool, Beyond emissions and economics: rethinking the co-benefits of electric vehicles (EVs) and vehicle-to-grid (V2G). Transp. Policy **71**, 130–137 (2018)
7. J. Tomić, W. Kempton, Using fleets of electric-drive vehicles for grid support. J. Power Sources **168**, 459–468 (2007)
8. S. Han, S. Han, K. Sezaki, Economic assessment on V2G frequency regulation regarding the battery degradation, in *2012 IEEE PES Innovative Smart Grid Technologies (ISGT)*, (Washington DC, 2012), pp. 1–6. https://doi.org/10.1109/ISGT.2012.6175717
9. A. De Los Rios, J. Goentzel, E. Nordstrom, C.W. Siegert, Economic analysis of vehicle-to-grid (V2G)- enabled fleets participating in the regulation service market, (Innovative Smart Grid Technologies (ISGT), IEEE, 2012)
10. A. Schuller, B. Dietz, C.M. Flath, C. Weinhardt, Charging strategies for battery electric vehicles: economic benchmark and V2G potential. IEEE Trans. Power Syst. **29**(5), 2014–2022 (2014). https://doi.org/10.1109/TPWRS.2014.2301024
11. L. Agarwal, W. Peng, L. Goel, Using EV battery packs for vehicle-to-grid applications: an economic analysis, in *2014 IEEE Innovative Smart Grid Technologies–Asia (ISGT ASIA)* (Kuala Lumpur, 2014), pp. 663–668
12. W. Zeng, J. Gibeau, M. Chow, Economic benefits of plug-in electric vehicles using V2G for grid performance-based regulation service, in *IECON 2015–41st Annual Conference of the IEEE Industrial Electronics Society*, (Yokohama, 2015), pp. 004322–004327. https://doi.org/10.1109/IECON.2015.7392772.
13. Y. Shirazi, E. Carr, L. Knapp, A cost-benefit analysis of alternatively fuelled buses with special considerations for V2G technology. Energy Policy **87**, 591–603 (2015)
14. D. Ciechanowicz, A. Knoll, P. Osswald, D. Pelzer, Towards a business case for vehicle-to-grid-maximizing profits in ancillary service markets, in A Book Chapter in Plug-in Electric Vehicles in Smart Grids—Energy Management (Springer, 2015), pp. 203–231
15. Y. Zhao, M. Noori, O. Tatari, Vehicle to grid regulation services of electric delivery trucks: economic and environmental benefit analysis. Appl. Energy **170**, 161–175 (2016)
16. C. Peng, J. Zou, L. Lian, L. Li, An optimal dispatching strategy for V2G aggregator participating in supplementary frequency regulation considering EV driving demand and aggregator's benefits. Appl. Energy **190**, 591–599 (2017)
17. Y. Kuang, Y. Chen, M. Hu, D. Yang, Influence analysis of driver behaviour and building category on economic performance of electric vehicle to grid and building integration. Appl. Energy **207**, 427–437 (2017)
18. R. Gough, C. Dickerson, P. Rowley, C. Walsh, Vehicle-to-grid feasibility: a techno-economic analysis of EV-based energy storage. Appl. Energy **192**, 12–23 (2017)
19. S. Tamura, T. Kikuchi, V2G strategy for frequency regulation based on economic evaluation considering EV battery longevity, in *2018 IEEE International Telecommunications Energy Conference (INTELEC)*, (Turin, 2018), pp. 1-6. https://doi.org/10.1109/INTLEC.2018.8612431
20. J. Liu, C. Zhong, An economic evaluation of the coordination between electric vehicle storage and distributed renewable energy. Energy **186**, (2019)

21. U. Datta, N. Saiprasad, A. Kalam, J. Shi, A. Zayegh, A price-regulated electric vehicle charge-discharge strategy for G2V, V2H, and V2G. Int. J. Energy Res. **43**, 1032–1042 (2019)
22. J. Lassila, J. Haakana, V. Tikka, J. Partanen, Methodology to analyze the economic effects of electric cars as energy storages. IEEE Trans. Smart Grid **3**(1), 506–516 (2012)
23. I.G. Unda, P. Papadopoulos, S.S. Kazakos, L.M. Cipcigan, N. Jenkins, E. Zabala, Management of electric vehicle battery charging in distribution networks with multiagent systems. Electr. Power Syst. Res. **110**, 172–179 (2014)
24. K. Mets, T. Verschueren, F. De Turck, C. Develder, Exploiting V2G to optimize residential energy consumption with electrical vehicle (dis)charging, in *2011 IEEE First International Workshop on Smart Grid Modeling and Simulation (SGMS)*, (Brussels, 2011), pp. 7–12.
25. T. Ma, O.A. Mohammed, Economic analysis of real-time large-scale PEVs network power flow control algorithm with the consideration of V2G services, in IEEE Transactions on Industry Applications, Vol. 50, No. 6 (2014)
26. K.N. Kumar, B. Sivaneasan, P.H. Cheah, P.L. So, D.Z.W. Wang, V2G capacity estimation using dynamic EV scheduling. IEEE Trans. Smart Grid **5**(2), 1051–1060 (2014)
27. M.J.E. Alam, K.M. Muttaqi, D. Sutanto, Effective utilization of available PEV battery capacity for mitigation of solar PV impact and grid support with integrated V2G functionality. IEEE Trans. Smart Grid **7**(3), 1562–1571 (2016)
28. K. Mets, T. Verschueren, W. Haerick, C. Develder, F. De Turck, Optimizing smart energy control strategies for plug-in hybrid electric vehicle charging, in *2010 IEEE/IFIP Network Operations and Management Symposium Workshops*, (Osaka, 2010), pp 293–299
29. Y. Mu, J. Wu, N. Jenkins, H. Jia, C. Wang, A spatial-temporal model for grid impact analysis of plug-in electric vehicles. Appl. Energy **114**, 456–465 (2014)
30. R. Lamedica, S. Teodori, G. Carbone, E. Santini, An energy management software for smart buildings with V2G and BESS. Sustain. Cities Soc. **19**, 173–183 (2015)
31. L. Drude, L.C. Pereira Junior, R. Rüther, Photovoltaics (PV) and electric vehicle-to-grid (V2G) strategies for peak demand reduction in urban regions in Brazil in a smart grid environment. Renew. Energy **68**, 443–451 (2014)
32. J.D. Fitzsimmons, et al., Simulation of an electric vehicle fleet to forecast availability of grid balancing resources, in *2016 IEEE Systems and Information Engineering Design Symposium (SIEDS)*, (Charlottesville, VA, 2016), pp. 205–210
33. A.Y.S. Lam, K.C. Leung, V.O.K. Li, Capacity estimation for vehicle-to-grid frequency regulation services with smart charging mechanism. IEEE Trans. Smart Grid **7**(1), 156–166 (2016)
34. Y. Wang, D. Infield, Markov Chain Monte Carlo simulation of electric vehicle use for network integration studies. Int. J. Electr. Power Energy Syst. **99**, 85–94 (2018)
35. Y. Wang, S. Huang, D. Infield, Investigation of the potential for electric vehicles to support the domestic peak load, *2014 IEEE International Electric Vehicle Conference (IEVC)*, (Florence, 2014), pp. 1–8
36. U.C. Chukwu, S.M. Mahajan, V2G parking lot with PV rooftop for capacity enhancement of a distribution system. IEEE Trans. Sustain. Energy **5**(1), 119–127 (2014)

Chapter 11
Communication and Networking Technologies in Internet of Vehicles

Yujie Song, Jianyong Song, Sihan Qin, and Yue Cao

Abstract Due to the rapid development of Intelligent Transportation System (ITS), application scenarios such as autonomous valet parking, convoy arrangement, vehicle–road cooperative control and vehicle emergency hedging have been realized. On the other hand, users need data application services with low latency, high quality and large data volume. So, communication and networking in the Internet of Vehicles (IoV) become the key to improve ITS system in the future. This chapter elaborates some different communication model in IoV. Then, each communication and networking model are proposed based on system model and confronting problems. In addition, this chapter introduces Vehicle-to-Vehicle (V2V), Vehicle-to-Infrastructure (V2I), Vehicle-to-Pedestrian (V2P), Vehicle-to-Network/Cloud (V2N/C), and trajectory-based communication modes. Finally, some use cases are discussed to list related work.

Keywords Message · Communication · Networking · IoV · V2V · V2I · V2P · V2N/C · Trajectory-Based

Abbreviations

ITS	Intelligent Transportation System
V2V	Vehicle-to-Vehicle
V2P	Vehicle-to-Pedestrian
V2X	Vehicles to Everything
RFID	Radio Frequency IDentification
LTE	Long Term Evolution
D-RAN	Distributed wiReless Access Network
TFB	Trajectory-Based Forwarding
BSMs	Basic Safety Messages

Y. Song (✉) · J. Song · S. Qin · Y. Cao
School of Cyber Science and Engineering, Wuhan University, Wuhan, China
e-mail: 120310899@qq.com

© The Author(s), under exclusive license to Springer Nature Singapore Pte Ltd. 2023
Y. Cao et al. (eds.), *Automated and Electric Vehicle: Design, Informatics and Sustainability*, Recent Advancements in Connected Autonomous Vehicle Technologies 3, https://doi.org/10.1007/978-981-19-5751-2_11

EFC	Electronic Fee Collection
TMS	Traffic Management Systems
RSU	Road Side Unit
DTNs	Delay/Disruption Tolerant Networks
VANET	Vehicular Ad hoc NETwork
IoV	Internet of Vehicles
V2I	Vehicle-to-Infrastructure
V2N	Vehicle-to-Network
D2D	Device to Device
IOT	Internet of Things
NB-IoT	Narrow Band Internet of Things
C-RAN	Centralized wiReless Access Network
DSRC	Dedicated Short Range Communication
NHTSA	National Highway Traffic Safety Administration
WCDMA	Wideband Code Division Multiple Access
VCPS	Vehicular Cyber-Physical Systems
RENA	REgioN-bAsed

11.1 Introduction

The vehicle–road collaboration of intelligent transportation and the autonomous driving of the automobile industry revolution are inseparable from the Internet of Vehicles. IoV refers to a communication network in which information is exchanged between "people, vehicles, roads and clouds" according to agreed communication protocols and data exchange standards. The industry in IoV is a new industry with deep integration of automobile, electronics, information and communication, road transportation and other industries. It is a hot spot of global innovation and a commanding height of future development.

Wireless communication technology for Internet of Vehicles (V2X, Vehicles to Everything) is a new generation of information and communication technology that enables all-round connection and communication between vehicles and surrounding cars, people, transportation infrastructure and cloud centers. What exactly is V2X feature vehicles? According to the China association of automobile manufacturers to carry V2X function definition of the car, it is carrying the advanced vehicle sensor, controller, actuator and other devices, and the integration of modern communication and network technology, realizes the car and X (people, vehicles, road, background, etc.) the exchange of intelligence information sharing, have complex environment perception, intelligent decision-making, coordination control and execution, and other functions, a new generation of vehicles capable of achieving safe, comfortable, energy saving, efficient driving and eventually replacing human operation. A new generation of vehicles capable of achieving safe, comfortable, energy saving, efficient driving and eventually replacing human operation.

11 Communication and Networking Technologies in Internet of Vehicles

As an extension of the application-oriented concept of IOT, V2X is an in-depth research process of D2D (Device to Device) technology. It refers to the communication system between vehicles, or between cars and pedestrians, cyclists and infrastructure. Using the load on the vehicle radio frequency identification (RFID) technology, sensors, cameras for vehicle traffic conditions, environmental information system running state and surrounding roads, at the same time with the aid of Global Positioning System (GPS) vehicle location information, and through the D2D technology will this information for end-to-end transmission, then realize the sharing of the information in the whole car networking system. Through the analysis and processing of these information, the road condition report and warning to the driver in time can effectively avoid the congested road and choose the best driving route.

V2X Internet of Vehicles communications are divided into four categories: Vehicle-to-Vehicle (V2V), Vehicle-to-Infrastructure (V2I), Vehicle-to-Person (V2P), vehicle-network/cloud and other communication modes. Among them, V2V, V2I and V2P have special communication conditions such as low delay, high reliability and dynamic connectivity, while V2N/C has slightly lower requirements. However, with the development of autonomous driving, assisted driving, human–machine cooperation and intelligent transportation system, IoV communication and IoV network have higher requirements for low delay, high load, high reliability, dynamic mobile networking, security and other characteristics.

For example, the perception of the surrounding environment by the unmanned driver requires rapid processing and sharing of the perceived data. The knowledge information exchange and calculation for the vehicle cluster require low delay and high computational capability. Therefore, various vehicle-based technologies require IoV communication to be efficient, reliable and safe.

At the same time, V2X communication also faces new scientific problems caused by vehicle movement, such as highly dynamic network topology and time–space complexity, complex and changeable radio transmission environment, low delay and high reliability communication problems under high density.

At present, China has elevated the Internet of Vehicles industry to a national strategic height, and the industrial policy continues to advance, and the industrial policy continues to be favorable, promoting the development of related technologies of the Internet of Vehicles.

The top-level design of the technical standard system of the Internet of Vehicles has been completed from the level of national standards. The industrialization process of the Internet of Vehicles in China is gradually accelerating, and a relatively complete industrial chain ecology has been formed around 5G-V2X, including communication chips, communication modules, on-board terminals, roadside equipment, vehicle manufacturing, operation services, testing and certification, high-precision positioning and map services.

11.2 Background

11.2.1 V2V Communication Models

V2V refers to vehicle-to-vehicle communication through vehicle-to-vehicle terminals. The most common application scenario is in the city streets and highways, vehicles communicate with each other, send data, and achieve information and data sharing. V2V technology allows vehicles to prevent accidents by forwarding real-time information about themselves and the road ahead, thereby reducing driving time and ultimately improving traffic conditions and reducing congestion. V2V vehicle communication is a commonly used model in the vehicle AD hoc network. Due to the high-speed mobility and uncertainty of vehicles, the topology of the whole network changes very fast, and the connection is unstable compared with the fixed network. Therefore, the V2V communication model mainly selects the forwarding vehicle as the main method, looks for the next hop vehicle, passes the message, and completes the purpose of information transmission.

Da Costa et al. [1] proposed a protocol, named DDRX, for Distributed Identification Algorithm of Data Transmission Relay Nodes Based on Complex Network Metrics. This protocol mainly relies on a storage-carry-forward mechanism, and considers the relative position and Angle of the forwarding vehicle and the source vehicle to ensure the transmission of messages. At the same time, it can effectively reduce the number of packets transmitted in the network and reduce the load of the network. Urmonov et al. [2] proposed an algorithm that can optimally forward vehicle sets under various network conditions, and a selective forwarding algorithm based on the transverse crossing (LCL) at traffic junctions. The algorithm proposed in this paper mainly focuses on identifying the best neighbor vehicle based on the basic information and preventing the message from traveling to the unexpected direction as much as possible. Cao et al. [3] proposed a track-based opportunistic routing protocol and designed a multi-queue system to prioritize messages with the highest delivery potential. The algorithm selects the forwarding vehicle according to the degree of vehicle trajectory fitting, reduces the load in the communication network, ensures a low level of average delivery delay, and selects the next hop forwarding vehicle by calculating the relationship between vehicle trajectory angles.

11.2.2 V2I Communication Models

It is a network in which nodes communicate with the roadside unit [4]. Due to the rapid mobility of vehicles, the time that vehicles are within range of each other does not support complete delivery of messages. In order to solve the problem of message delivery within a limited time, many scholars consider the method of V2I assisted message transmission to solve the problem of message delivery. Therefore, in terms of V2I information acquisition, a set of complete traffic condition information collection method based on floating vehicle data and a set of traffic information expression system based on road chain are proposed, and the first practical floating vehicle processing system in China and the largest in the world are built [5].

Vijayakumar et al. [6] proposed Adaptive Load Balancing Schema for efficient data dissemination in Vehicular Ad-Hoc Network VANET. The authors established an effective vehicle data transmission model in a collaborative manner with available RSUs. The model considers real time delay and delay free tolerance, providing an efficient communication method to make the data flow between vehicles large and jitter free. Abhilasha Sharma et al. [7] proposed Adaptive Priority Scheduling (AdPS) for Data Services in Heterogeneous Vehicular Networks. This paper uses fuzzy request deadline estimation model, fuzzy first search model, and an adaptive data service scheduling algorithm to evaluate the final time of requests given real-time vehicle information, calculate priorities based on real-time vehicle and service type information, and effective real-time data service scheduling, respectively. Liu et al. [8] proposed Temporal Data Dissemination in Vehicular Cyber-physical Systems. This paper proves that the temporary data propagation formula is NP-hard through the clustering problem, and proposes a heuristic scheduling algorithm to improve overall system performance through broadcast efficiency, bandwidth utilization, and request service opportunity.

11.2.3 V2P Communication Models

Vehicles to Pedestrian (V2P) refers to the communication between vehicle-mounted devices and vulnerable traffic groups (including pedestrians, cyclists, etc.) using user devices (including pedestrians, cyclists, etc.) and user devices (such as smart phones, wearable devices, bicycle GPS signal devices). Vehicular and pedestrian communication is mainly used in traffic safety, intelligent keys, location information services, car sharing, etc. Using smartphones or wearable devices to detect the position of pedestrians, direction and speed. Shortwave communication technology is used to obtain the position, direction and speed of surrounding vehicles. Then, the system calculates the motion state and notifies the pedestrian and driver if a collision is possible, or avoids the accident through the vehicle's control obstacle avoidance system. V2P enables pedestrians and cyclists to become nodes in the V2X communication environment via their smartphones. It can send or receive warning signals, for example, it can inform the network signal lamp in advance whether it needs to extend the time to cross the road, it can also indicate that there is a pedestrian to cross the road in front of the nearby vehicle, or there is a bicycle riding in the adjacent lane of the vehicle. V2P communication mainly depends on the wireless communication protocol of the user's device, which requires the intelligent vehicle computing platform to support as many wireless communication protocols as possible, such as BT, WIFI, Zigbee, Z-Wave, NB-IOT /LoRA, LTE/5G, to effectively implement V2P.

Mehrdad Bagheri et al. [9] proposed an approach that utilizes cellular technologies. This paper shows potential of utilizing 3G and LTE for highly mobile entities of vehicular network applications. The Author adopts an adaptive multistage approach that runs in energy-saving mode in a risk-free situation but switches to normal mode when risky situations are detected. Ahmed Hussein et al. [10] proposed based on Pedestrian to Vehicle (P2V) and Vehicle to Pedestrian (V2P) communication technologies, which increases the visual situational awareness of VRU regarding the nearby location of both autonomous and manually-controlled vehicles in a user-friendly form.

11.2.4   V2N/C Communication Models

Vehicle to Network/Cloud (V2N/C) refers to the connection between the on-board equipment in the vehicle and the cloud platform through the network, data interaction between the cloud platform and the vehicle, storage and processing of the obtained data, and the provision of remote traffic information push, entertainment, business services and vehicle management.

Communication between vehicle and cloud platform is mainly used in vehicle navigation, vehicle remote monitoring, emergency rescue, infotainment services, etc. For V2N/C, the intelligent vehicle computing platform requires powerful and fast data processing capabilities and massive data storage mechanisms to process network data with ultra-high speed, ultra-high throughput, high reliability, and ultra-low latency.

V2N/C will need to use c-V2X's new antenna transmission technology, high frequency transmission, and full duplex in the same frequency. This requires the intelligent vehicle computing platform to support new network architectures, such as D-RAN, C-RAN, D2D, and MTC access technologies, and a variety of low cost and high-speed optical transmission networks.

Kohei Moto et al. [11] worked on research and development of truck platooning as a new 5G application. The authors have developed a field trial system for vehicular-to-network (V2N) communications using 5G prototype equipment and actual large-size trucks in order to assess 5G capabilities, including ultra-low-latency, in automotive test courses in the field. Furthermore, the paper addresses the field evaluation results of 5G V2N communications in a rural area.

11.2.5   Trajectory-Based Communication Models

Routing packets along a specified curve is a new approach to forwarding packets in large scale dense ad-hoc networks. The fundamental aspect of considering route paths as continuous functions is the decoupling of the path name from the path itself [12]. Trajectory-Based Forwarding (TBF) provides a common framework for many services such as: broadcasting, discovery, unicast, multicast and multipath routing in ad hoc networks [13]. TBF is a novel method to forward packets in a dense ad hoc network that makes it possible to route a packet along a predefined curve. It is a hybrid between source based routing and Cartesian forwarding in that the trajectory is set by the source, but the forwarding decision is based on the relationship to the trajectory rather than names of intermediate nodes [14]. There are essential differences between trajectory-based communication mode and other communication models. TBF has two different forwarding categories: one is to forward messages according to the reference trajectory, and the other is to forward messages according to the reference curve. However, its essence is to forward messages with different strategies according to a known trajectory. The reference trajectory has an important influence on message forwarding. For example, the linear trajectory between the source node and the destination node will reduce the number of intermediate forwarding nodes. If a more tortuous trajectory is adopted, the number of intermediate forwarding nodes will be increased. Therefore, trajectory generation has an important influence on the trajectory-based communication models.

Cao et al. [15] proposed The Best Geographic Relay (TBGR) routing scheme to relay messages via a limited number of copies, under the homogeneous scenario. Wen et al. [16] proposed a storage-friendly REgioN-bAsed protocol (RENA) which builds routing tables based on regional movement history, which avoids excessive storage for tracking encounter history. Murat Yuksel et al. [14] proposed techniques to efficiently forward packets along a trajectory defined as a parametric curve which is the well-known Bezier parametric curve for encoding trajectories into packets at source.

11.2.6 Technical Standard for V2X

11.2.6.1 Dedicated Short Range Communication (DSRC)

In the United States and many other countries, parties are actively developing and deploying DSRC. In February 2014, the U.S. Department of Transportation first pledged its strong support for DSRC applications in light vehicles. Since then, the U.S. Congress, the U.S. Department of Transportation, the IEEE, and major automakers have actively recommended DSRC legislation. It should be noted that the National Highway Traffic Safety Administration (NHTSA) requires DSRC as the safety standard for V2V vehicles in the near future. The vehicle will send and receive Basic Safety Messages (BSMs) via DSRC. This requirement has been overwhelmingly supported by the car enterprises. Only a handful of companies in the mobile and Wi-Fi sectors have objected. The timetable for large-scale DSRC deployment is shown in Fig. 11.1.

Dedicated Short Range Communication (DSRC) is a technology in the field of ITS intelligent transportation system, which is specially used for motor vehicles to realize automatic charging Electronic Fee Collection (EFC) without stopping at the toll points such as highways. It's called long distance RFID, or E-Tag. DSRC is

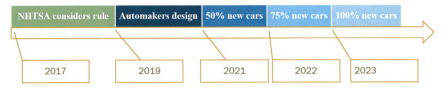

Fig. 11.1 The timetable for large-scale DSRC deployment

dedicated to ETC, but its communication mode can be applied to the Internet of vehicles. This is because vehicles are on the road, which is generally dense, and can communicate with each other through on-board units and DSRC technology.

The target communication range of DSRC is within 1 km. Compared with cellular and satellite communication, the communication distance is shorter. The advantage of DSRC is that it is currently in use in many countries and the technology is mature. The disadvantage is that the transmission distance is short, the signal is easy to be blocked by buildings, and many signal transmitters need to be built repeatedly.

11.2.6.2 Celluler-V2X (C-V2X)

Long Term Evolution V2X (LTE-V2X)

LTE-V2X is mainly aimed at V2X basic road safety applications, and 3GPP includes multiple use cases including vehicle-vehicle, vehicle–road, vehicle-person and vehicle-background communication, including forward collision prevention, vehicle loss of control alarm, etc. The main challenges for LTE-V2X are the low latency, high reliability, large capacity, high communication frequency, high density communication scenarios and performance requirements proposed by vehicle-to-vehicle and vehicle-to-road communications. LTE-V2X provides two modes of communication: cellular (also known as LTE-V-Cell) and Direct (also known as LTE-V-Direct), and the two modes complement each other. In the cellular network coverage, terminals can intelligently select the best transmission channel, and the cellular network can carry out various forms of direct link management and configuration. In scenarios without cellular network coverage, the direct mode can still work independently, effectively ensuring the continuity of IoV services.

Centralization model (LTE-V-Cell)

Using the cellular network as a centralized control center and data information forwarding center, centralized scheduling, congestion control and interference coordination can significantly improve LTE-V2X access and networking efficiency, and ensure service continuity and reliability. The Uu interface which corresponds to the

11 Communication and Networking Technologies in Internet of Vehicles

cellular communication mode is the fixed network part interface for the user equipment to access the Wideband Code Division Multiple Access (WCDMA) system. LTE-V2X enhances the Uu interface to better support V2X services.

Distributed model (LTE-V-Direct)

Aiming at the challenges of road safety services such as low latency, high reliability transmission requirements, high-speed node movement and hidden terminals, LTE-V2X proposes the physical layer technology, resource allocation method, quality of service management and congestion control mechanism of LTE-V2X direct link. Direct communication can work in and out of cell coverage. When working within cellular coverage, resource allocation can be controlled centrally by the base station or distributed. Distributed control is used when working outside the cell coverage. Therefore, direct communication mode (LTE-V-Direct) is also called distributed model.

5g-V2x(Nr-V2x)

NR-V2X is an important component of C-V2X to support the communications needs of future internet-of-vehicles enhanced applications. Vehicle formation, advanced driving, sensor extension and remote driving are four types of IoT applications and related communication requirements, which require more reliable, lower latency and higher data rate IoT communication services over the direct link, with minimum end-to-end latency of 3 ms and reliability of 99.999%. The data rate of the direct link must be up to 1Gbit/s, which imposes more stringent requirements on C-V2X technology evolution [17].

To ensure the orderly development of Internet of Vehicles technology and industry, the research and standardization process of NR-V2X includes the design of LT-V2X coexistence. At the same time, it is emphasized that the relationship between LT-V2X and NR-V2X is complementary rather than substitute [18, 19]. LTE-V2X is used for V2X basic road safety services. NR-V2X is geared toward enhanced V2X business.

Considering the actual deployment of C-V2X, lT-V2X and NR-V2X will coexist for at least a period of time. Cellular networks will be able to control LTE-V2X and NR-V2X communications when LTE-V2X and NR-V2X operate within the range of cellular networks. The 5G access network includes two types of nodes: 5G NR system base station (gNB, Next Generation NodeB) and LTE 4G evolution base station (ng-eNB, Next Generation eNB) which can access the 5G core network.

The architecture of 5G-V2X(NR-V2X)

The NR-V2X network architecture should be designed to flexibly support the above two types of coverage. For details, we can read the standard 3GPP TS 23.287 (2020). Figure 11.2 shows the reference network architecture of NR-V2X, and Table 11.1

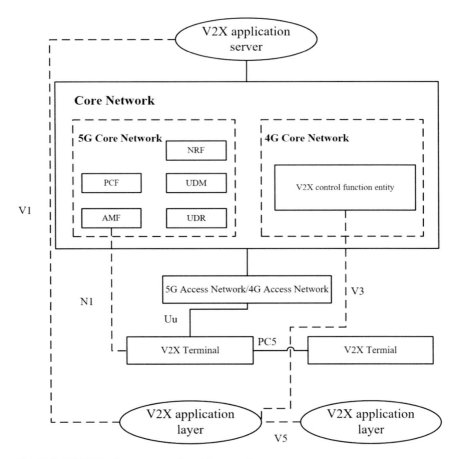

Fig. 11.2 NR-V2X reference network architecture diagram

describes the corresponding reference ports. Contract information: Contract information signed between a business user or business provider and a business or network operator. Authentication information used to verify whether a user has the permission to access the system.

- V2X Application Server

    V2X management entities outside the cellular network can manage global V2X communication policies and parameters, and V2X terminal signing and authentication information.

- 4G Core Network (4GC)

    The 4GC connects to the V2X application server and manages V2X communication policies and parameter configurations, contract information, and authentication information for V2X terminals in cellular coverage through the V2X control function (VCF) entity.

Table 11.1 Common communication technology

	DSRC	C-V2X	
		LTE-V2X	5G-V2X
Constitutor	IEEE	3GPP	3GPP
Supporter	U.S Department of Transportation	Some chip manufacturers, communication enterprises	Some chip manufacturers, communication enterprises
Represent the enterprise	NXP semiconductors	Qualcomm, Hua Wei	Hua Wei, Qualcomm
Normal condition	The standardization process began in 2004. Now that's done	The standardization process began in 2014. Now that's done	The standardization process began in 2016
Average latency	Low(<50 ms)	High(>50 ms)	Lower(<3 ms)
Bandwidth	High	Higher	Higher

- 5G Core Network (5GC)

  5GC connects to a V2X application server to configure V2X communication policies and parameters, and manage contract information and authentication information for V2X terminals on the cellular coverage network. Different from 4GC, 5GC adopts a service-based architecture, and each network function can evolve and expand independently. 3GPP puts the V2X Control Function (CVF) in 4GC in the Policy Control Function (PCF) in 4GC in NR-V2X. Accordingly, the functional entities of 5G core network are expanded:

  - Unified Data Repository (UDR) function extension.
  - Unified Data Management (UDM) function extension.
  - Policy Control Function (PCF) extension.
  - Access and Mobile Management Functions (AMF) extension.
  - Network Repository Function (NPF) extension.

- V2X Terminal

  Based on the obtained V2X communication policies and parameter configurations of PC5 and Uu, V2X communication is implemented on the PC5 or Uu ports.

  NR-V2X Terminal Authorization, Policy, and Parameter Configuration Procedure Reference (3GPP TS 38.300, v16.1.0). When the terminal is within the coverage of the cellular network, the corresponding authorization and parameter configuration are carried out over the network. When a terminal is located outside a cell, the terminal can communicate with each other over a direct link only when it knows its location information. NR-V2X can be configured with different PC5 wireless access technologies for different V2X applications, thus achieving the complementary relationship between LTE-V2X and NR-V2X.

Table 11.2 NR-V2X reference interface

Reference interface	Function
V1	Reference interface between the terminal V2X application and the V2X application server
V3	On a 4G network, a reference interface between a V2X application terminal and a V2X control entity (VCF, V2X Control Function)
V5	Reference interface between different V2X applications
N1	Reference ports for 5G networks from access and mobility management functions (AMF, Access and Mobility Management Function) to V2X terminals

V2X terminals can obtain V2X communication authentication, signing, policy configuration, and parameter configuration information in four ways. In descending order of priority as follow:

- V2X configuration information obtained through PCF (N1 reference interface).
- V2X configuration information obtained from a V2X application server (V1 reference port).
- V2X configuration information is obtained from information stored in the USIM (Universal Subscriber Identity Module).
- Obtain the V2X configuration information from the preconfigured information on the device.

As the first phase of C-V2X research and standardization, LTE-V2X provides the first opportunity to combine cellular and direct communication technologies. It is designed to optimize and enhance the system architecture, wireless air ports, and other areas based on LTE systems to effectively support basic road safety services with low latency and high reliability. The innovative design of the direct link for periodic and small data packet broadcasting is carried out, and the corresponding technical principles such as the physical layer channel structure, resource allocation method and synchronization mechanism have become the important foundation of NR-V2X technology. NR-V2X extends and enhances these technologies. As the second phase of C-V2X research and standardization, NR-V2X is designed to support communications requirements for enhanced IoV applications (Table 11.2).

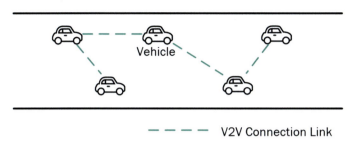

Fig. 11.3 V2V communication model

11.3 Use Cases

11.3.1 V2V Communication Use Cases

11.3.1.1 System Model

System framework is shown in Fig. 11.3, which is mainly including vehicles and road. Each vehicle is equipped with Global Position System (GPS) and captures its won movement information, including current location, moving direction and speed [3]. Vehicles can communicate with each other within the communication range by establishing links. When the vehicle is out of communication range, the link is disconnected.

11.3.1.2 Confronting Problem

When the neighbor vehicle is no longer the final receiving vehicle of the message, we complete the message transmission through the form of relay. Many of these vehicular applications require efficient data dissemination schemes for their realization [20]. For example, traffic status warning, fleet arrangement, vehicle coordinated scheduling, autonomous passenger parking, intelligent parking, etc. However, V2V communication also faces the following challenges:

- Broadcast storm: there are some V2V message transmission mechanisms that do not control the number of message copies. When a vehicle copies a message, a large number of message copies exist in Ad hoc Networks.
- High end-to-end latency: Due to the mobility of vehicles, messages have a probability of reverse propagation, resulting in increased delay of message transmission to the destination.

11.3.1.3 Related Work

Because the Traffic Management Systems (TMS) services request data, communication, and processing operations. In addition, VANET allows direct communication between direct vehicles, and data can be exchanged and processed between vehicles, so vehicle Ad hoc Networks have a strong impact on TMS applications. In TMS services, data transfer between vehicles is required, and a large amount of data flooding the vehicle Ad hoc Networks will pose a network load challenge. Due to the characteristics of the vehicle Ad hoc Networks, such as short-range communication and high mobility nodes, this will lead to topology variability.

Da Costa et al. [1] introduced an extensive analysis of data dissemination protocol based on complex network's metrics for urban VANET scenarios, called DDRX. Each vehicle must build a subgraph to identify the relay node to continue the dissemination process. Based on the local graph, it is possible to select the relay nodes based on complex networks' metrics [1]. The distributed identification algorithm and the operation of DDRX protocol for data transmission relay nodes based on complex network metrics are described in detail. Finally, the authors extended estimation in more realistic scenarios for data propagation and traffic management, including new scenarios, metrics, and evaluation protocols. Urmonov et al. [2] proposed an efficient message broadcast propagation method based on Lateral Crossing Line (LCL) algorithm. Based on the position of vehicles on the road, the method calculates an area to select a set of candidate vehicles. The author selects the next hop vehicle by calculating the propagation time of the message in this area. The smaller the transmission time, the more likely the vehicle will become a forwarding vehicle. Cao et al. [3] proposed a Trajectory-Driven Opportunistic Routing (TDOR) protocol, which is primarily applied for sparse networks, e.g., Delay/Disruption Tolerant Networks (DTNs). When the geographic routing protocol is involved in DTN, the existing work mainly considers the proximity to the destination as the standard selection of the next hop.

11.3.2 V2I Communication Use Cases

11.3.2.1 System Model

System framework is shown in Fig. 11.4, which is mainly including vehicles and Road Side Units (RSU). Road side units are deployed along with road at a certain location, and behaves as broker to bridge the information flow from vehicles to routers. Each RSU is able to aggregate all vehicles condition information and caches it in local storage [21]. When the vehicle is within the coverage area of the RSU, the vehicle establishes a connection with the RSU. Once the vehicle leaves the communication range of RSU, the connection is interrupted.

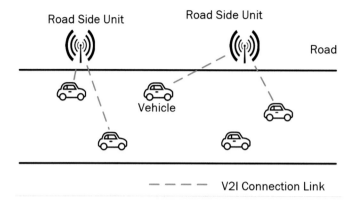

Fig. 11.4 V2I communication model

11.3.2.2 Confronting Problem

Since the RSUs are fixed in position, the vehicle can only connect to the RSUs within the communication range. The government needs to build road infrastructure and assemble RSUs along the roads. This is bound to be a long process. In order to ensure efficient communication between vehicles and RSU, scholars need to propose more efficient communication strategies or algorithms. RSU needs to request, broadcast and process important information and exchange it between vehicles in the proper time [4]. To be sure, the computing and storage capabilities of RSUs have also been challenged.

11.3.2.3 Related Work

In V2I communication, RSU aggregates a large number of messages. Therefore, messages need to be processed in the RSU (for example, merge, split, sort, etc.). Between RSUs, the overall network load needs to be considered. Vijayakumar et al. [6] proposed work focuses on efficient data dissemination on V2I communication model. This model considers real world time delay without imposing delay tolerance. This paper improved the existing cooperative load balancing (CLB) method to solve the case where the data transfer time is less than the remaining time of the vehicle in the RSU and the actual delay is too high. Adaptive load balancing model for V2I data delivering adopts different real-time scheduling algorithms for the priority of the requests to be processed. To improve the real-time information service capability in the Internet of vehicles, there are still many challenges according to the characteristics of the network. For efficient information dissemination, there is a prime requirement of an adaptive scheduling algorithm for approximate vehicular dynamics information [7]. Abhilasha Sharma et al. [7] proposed a novel Adaptive priority data Service scheduling (AdPS) algorithm to provide real-time data services, which is

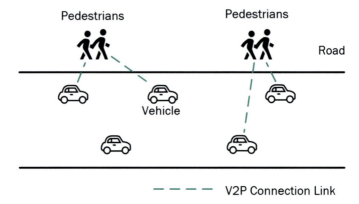

Fig. 11.5 V2P communication model

based on fuzzy logic request deadline estimation and request prioritization. The author proposes two models and an algorithm, namely Fuzzy Request Deadline Estimator Model, Fuzzy Priority Index Model and Adaptive Data Services Scheduling algorithm respectively. Liu et al. [8] presented the first study on real-time data services via roadside-to-vehicle communication by considering both the time constraint of data dissemination and the freshness of data items. In this paper, the authors examined some problems, for example, strict time limits for service requests, a request may require multiple related data items, and the query cannot be fully processed until all requested data items are retrieved, etc.

11.3.3 V2P Communication Use Cases

11.3.3.1 System Model

System framework is shown in Fig. 11.5, which is mainly including vehicles and pedestrians. Vehicles and pedestrians will be in communication with each other and will be connected. When either party leaves the communication range of the other, the connection is disconnected.

11.3.3.2 Confronting Problem

V2P communication has the following applications: emergency hedging, emergency message dissemination, small edge computing and so on. Due to the limited power of pedestrians' mobile phones or computers, vehicles move relatively fast, making it impossible to ensure an effective connection for a long time. Therefore, V2P communication time is short. In an emergency, for example, pedestrians may not have much time to react the emergency status. The limited computing power and time of mobile phone or computer becomes the challenge of small edge computing.

11.3.3.3 Related Work

Road safety is one of the most important applications of vehicle network. However, improving pedestrian safety through vehicle-to-pedestrian (V2P) wireless communication has not been widely addressed [9]. The communication application between people and vehicles has an important role in ensuring the area of road safety. From the previous literature, we can infer that the pedestrian safety application is time sensitive; hence, the time restraints are very tight and a complete investigation of latency in diverse P2I/V2P communication should be examined and assessed [4]. In this article [9], the authors' vision is to present an approach that enables the development of V2P road safety applications over wireless communication and using only existing infrastructure and equipment. The authors vary the frequency of message delivery according to the danger level of the event, while considering the energy consumption of mobile phones. Finally, the results show that this adaptive method saves a lot of electrical energy, which makes the road safety system based on cellular feasible. Ito et al. [22] proposed proximity relaying to address channel congestion problems which due to more pedestrian terminals such as smartphones and increased battery consumption. Two challenges remain in V2P communication, one is low latency V2P message transmission and the other is effective reduction of user power consumption. To solve the above challenges, Li et al. [23]proposed a novel solution, V2PSense, which trades off positioning precision for energy savings while achieving low-latency message transport with LTE high-priority bearers. The method proposed by the author can ensure the danger point near pedestrians even when GPS positioning is not particularly accurate.

11.3.4 V2N Communication Use Cases

11.3.4.1 System Model

The system framework is shown in Fig. 11.6, which is mainly including vehicles, edge clouds and cloud center. According to the application requirements of V2N, the communication between vehicles and the cloud needs to be realized through the base station (BS). BS realizes the function of edge cloud and is responsible for message processing, forwarding, computing, information sharing and other functions.

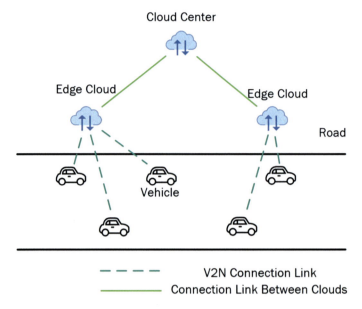

Fig. 11.6 V2N communication model

11.3.4.2 Confronting Problem

V2N implements large-range connection, multi-user access, and high-efficiency message transmission, processing, and computing. Edge cloud provides application data service for users, so it needs to process and transmit a large amount of data. In the communication of Internet of vehicles, the delay requirement of message is low, so how to make the message more efficient, stable and safe transmission is the problem that needs to be solved today.

11.3.4.3 Related Work

Cellular V2N communication will be backbone of the connected vehicles of future [24]. Vehicular-to-Network (V2N) means the communication between a car and an application service/server [25]. V2N is mainly used to share information between vehicles and the cloud. Vehicles can send vehicle and traffic information to the cloud traffic police command center, and the cloud can also send broadcast information, such as traffic congestion and accidents, to relevant vehicles in a certain area. V2V and V2I represent near field communication, while V2N technology enables remote data transmission. Kohei Moto et al. [11] worked on the research and development of truck platooning as a new 5G application, and evaluated the radio transmission environment of V2N. This paper addresses the field evaluation results of 5G V2N communications in a rural area. V2N provides services for a large number of users.

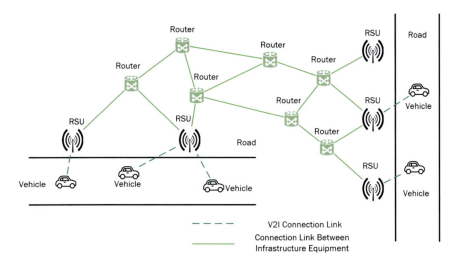

Fig. 11.7 Trajectory based communication model

Therefore, application services provided by V2N are protected from attacks. Xu et al. [26] proposed a novel hash coding method to probabilistically encode/decode random pilots using resource features of wireless data, such that the influence of attack on pilots can be dispersed across resources and minimized under arbitrary distribution of attack modes.

11.3.5 Trajectory Based Communication Use Cases

11.3.5.1 System Model

System framework is shown in Fig. 11.7, which is mainly including vehicles, RSUs and routers. RSUs is responsible for receiving, forwarding, and setting the initial trajectory for messages. Meanwhile, track-based message forwarding algorithms are also used in Routers and RSUs. As a relay device, routers are designed to transfer messages faster.

11.3.5.2 Confronting Problem

Due to high mobility and sparse network density, Vehicular Cyber-Physical Systems (VCPS) could be severely affected by intermittent connectivity [3]. The uncertainty of environment usually requires disposing hundreds of sensors to cooperate, and the research to sensor network is regarded as a challenging area especially when considering its characteristics of high density of nodes and limited resource of nodes

[27]. At present, the rapid development of big data not only faces the challenge of a large number of data packets, but also faces great challenges in infrastructure allocation and message forwarding algorithms.

11.3.5.3 Related Work

Previous geographic routing schemes in Delay/Disruption Tolerant Networks (DTNs) only consider the homogeneous scenario where nodal mobility is identical [15]. To fill this gap, they first proposed TBGR routing scheme to forward messages through limited copies of messages. Authors further overcome the problem of local maximization in the case of sparse network density with TBGR, different from those efforts in dense networks, for example, clustered wireless sensor networks. Then, they extended the TBGR heterogeneous scenario and proposed the TBHGR routing mechanism. On this basis, the authors consider a single node access reference (referred to nonidentical nodal mobility). The author selects the next hop vehicle forwarding by taking the running angles and vehicle categories between vehicle tracks as reference conditions. Considering the constraints brought by mobility and resources, it is important for routing protocols to efficiently deliver data in Intermittently Connected Mobile Network (ICMN). Wen et al. [16] proposed a storage-friendly REgioN-bAsed protocol (RENA) for efficiently deliver data in ICMN. Instead of using temporal information, RENA builds routing tables based on regional movement history, which avoid excessive storage for tracking encounter history. By calculating the message delivery time between regions, the author calculates the return time combined with the vehicle trajectory, and then selects the optimal message delivery path.

11.4 Conclusion

This chapter introduced various communication models in Intelligent Transportation System (ITS), including V2V, V2I, V2P, V2N/C, and trajectory-based communication models. Then, numerals use cases were introduced for more details on the communication models. To support the application scenarios of ITS, such as, autonomous valet parking, convoy arrangement, vehicle–road cooperative control, and vehicle emergency hedging, the above works have provided a variety of communication networking models. There are five categories of communication models in ITS, i.e., V2V, V2I, V2P, V2N/C, and trajectory-based communication models. Furthermore, we exposed the key issues faced by each communication model, introducing the concepts, characteristics, and system models of each communication model. Finally, all the work is to improve the message delivery ratio, and reduce the delivery latency and overhead ratio.

References

1. J.B. da Costa, A.M. de Souza, D. Rosário, E. Cerqueira, L.A. Villas, Efficient data dissemination protocol based on complex networks' metrics for urban vehicular networks. J. Internet Serv. Appl. **10**(1), 1–13 (2019)
2. O. Urmonov, H. Kim, A multi-hop data dissemination algorithm for vehicular communication. Computers, **9**(2), 25 (2020)
3. Y. Cao, O. Kaiwartya, N. Aslam, C. Han, X. Zhang, Y. Zhuang, M. Dianati, A trajectory-driven opportunistic routing protocol for VCPS. IEEE Trans. Aerosp. Electron. Syst. **54**(6), 2628–2642 (2018)
4. P. Sewalkar, J. Seitz, Vehicle-to-pedestrian communication for vulnerable road users: survey, design considerations, and challenges. Sensors **19**(2), 358 (2019)
5. W. He, H. Li, X. Zhi, X. Li, J. Zhang, Q. Hou, Y. Li, Overview of V2V and V2I wireless communication for cooperative vehicle infrastructure systems, in *2019 IEEE 4th Advanced Information Technology, Electronic and Automation Control Conference (IAEAC)*, (IEEE, Chengdu, China, 2019), pp. 127–134
6. V. Vijayakumar, K.S. Joseph, Adaptive load balancing schema for efficient data dissemination in vehicular Ad-Hoc network VANET. Alex. Eng. J. **58**(4), 1157–1166 (2019)
7. A. Sharma, L.K. Awasthi, AdPS: adaptive priority scheduling for data services in heterogeneous vehicular networks. Comput. Commun. **159**, 71–82 (2020)
8. K. Liu, V.C.S. Lee, J.K.Y. Ng, J. Chen, S.H. Son, Temporal data dissemination in vehicular cyber–physical systems. IEEE Trans. Intell. Transp. Syst. **15**(6), 2419–2431 (2014)
9. M. Bagheri, M. Siekkinen, J.K. Nurminen, Cellular-based vehicle to pedestrian (V2P) adaptive communication for collision avoidance, in *Proceeding of the 2014 international conference on connected vehicles and expo (ICCVE)*, (IEEE, Vienna, Austria, 2014), pp. 450–456
10. A. Hussein, F. Garcia, J.M. Armingol, C. Olaverri-Monreal, P2V and V2P communication for pedestrian warning on the basis of autonomous vehicles. in *2016 IEEE 19th International Conference on Intelligent Transportation Systems (ITSC)*, (IEEE, Rio de Janeiro, Brazil, 2016), pp. 2034–2039
11. K. Moto, M. Mikami, K. Serizawa, H. Yoshino, Field experimental evaluation on 5G V2N low latency communication for application to truck platooning, in *2019 IEEE 90th Vehicular Technology Conference (VTC2019-Fall)*, (IEEE, Honolulu, HI, USA, 2019), pp. 1–5
12. B. Nath, D. Niculescu, Routing on a curve. ACM SIGCOMM Comp Commun Rev **33**(1), 155–160 (2003)
13. H. Houda, M. Salah, A survey of trajectory based data forwarding schemes for vehicular ad-hoc networks, in *2015 IEEE International Conference on Communication Software and Networks (ICCSN)*, (IEEE, Chengdu, China, 2015), pp. 399–404
14. M. Yuksel, R. Pradhan, S. Kalyanaraman, Trajectory-based forwarding mechanisms for ad-hoc sensor networks, in *Proc. of the IEEE 2nd Upstate New York Workshop on Sensor Networks*, (Syracuse, NY, 2003)
15. Y. Cao, K. Wei, G. Min, J. Weng, X. Yang, Z. Sun, A geographic multicopy routing scheme for DTNs with heterogeneous mobility. IEEE Syst. J. **12**(1), 790–801 (2016)
16. H. Wen, J. Liu, C. Lin, F. Ren, P. Li, Y. Fang, Rena: region-based routing in intermittently connected mobile network, in *Proceedings of the 12th ACM international conference on Modeling, analysis and simulation of wireless and mobile systems*, (Association for Computing Machinery, Tenerife, Canary Islands Spain, 2009), pp. 280–287
17. S. Chen, J. Hu, Y. Shi, L. Zhao, W. Li, A vision of C-V2X: technologies, field testing, and challenges with chinese development. IEEE Internet Things J. **7**(5), 3872–3881 (2020)
18. 3GPP TS 23.287, v16.1.0. Architecture enhancements for 5G system (5GS) to support vehicle-to-everything (V2X) services (Release 16) [S]. (2020)
19. 3GPP TS 38.300, v16.1.0. NR and NG-RAN overall description[S] (2020)
20. M. Laha, R. Datta, Efficient message dissemination in V2V network: a local centrality-based approach, in *2021 National Conference on Communications (NCC)*, (IEEE, Kanpur, India, 2021), pp. 1–6

21. Y. Cao, S. Yang, G. Min, X. Zhang, H. Song, O. Kaiwartya, N. Aslam, A cost-efficient communication framework for battery-switch-based electric vehicle charging. IEEE Commun. Mag. **55**(5), 162–169 (2017)
22. H. Ito, T. Murase, K. Sasajima, Congestion control and energy savings for V2P communication crash warnings with proximity relying, in *2016 International Conference on Connected Vehicles and Expo (ICCVE)*, (IEEE, Seattle, WA, USA, 2016), pp. 43–48
23. C.Y. Li, G. Salinas, P.H. Huang, G.H. Tu, G.H. Hsu, T.Y. Hsieh, V2PSense: enabling cellular-based V2P collision warning service through mobile sensing, in *2018 IEEE International Conference on Communications (ICC)*, (IEEE, Kansas City, MO, USA, 2018), pp. 1–6
24. U. Saeed, J. Hämäläinen, E. Mutafungwa, R. Wichman, D. González, M. Garcia-Lozano, Route-based radio coverage analysis of cellular network deployments for V2N communication, in *2019 International Conference on Wireless and Mobile Computing, Networking and Communications (WiMob)*, (IEEE, Barcelona, Spain, 2019), pp. 1–6
25. T. Deinlein, R. German, A. Djanatliev, 5G-Sim-V2I/N: towards a simulation framework for the evaluation of 5G V2I/V2N use cases, in *2020 European Conference on Networks and Communications (EuCNC)*, (IEEE, Dubrovnik, Croatia, 2020), pp. 353–357
26. D. Xu, J.A. Ritcey, Hashed anti-denial of access for V2N URLLC services: a probabilistic graph approach. IEEE Trans. Veh. Technol. **70**(10), 10077–10092 (2021)
27. J. Zhang, Y.P. Lin, M. Lin, P. Li, S.W. Zhou, Curve-based greedy routing algorithm for sensor networks, in *International Conference on Networking and Mobile Computing*, (Springer, Berlin, Heidelberg, 2005), pp. 1125–1133
28. S.A. Ahmad, M. Shcherbakov, A survey on routing protocols in vehicular adhoc networks, in *2018 9th International Conference on Information, Intelligence, Systems and Applications (IISA)*, (IEEE, Zakynthos, Greece, 2018), pp. 1–8

Chapter 12
The Overview of Non-orthogonal Multiple Access in Vehicle-to-Vehicle Communication

Lei Wen and Yue Cao

Abstract Non-orthogonal multiple access (NOMA) is an emerging technology for vehicle-to-vehicle (V2V) communication in order to accommodate more users via non-orthogonal resource allocation, especially when the number of users/devices exceeds the available degrees of freedom, resulting in an overloaded condition. To date, NOMA is a promising technique in vehicle-to-vehicle communication. However, research in this area is still in its infancy and there are still several open issues in the transceiver design. The goal of this chapter is to bring this upcoming trend to the attention of the communications communities and to motivate more research in this important area.

Abbreviations

2G	The Second Generation
3G	The Third Generation
4G	The Forth Generation
5G	The Fifth Generation
AWGN	Additive White Gaussian Noise
BER	Bit Error Rate
CDMA	Code Division Multiple Access
CND	Check Node Detector
CP	Cyclic Prefix
EXIT	Extrinsic Information Transfer
FEC	Forward Error Correction

L. Wen (✉)
College of Electronic Science and Technology, National University of Defense Technology, Changsha, China
e-mail: newton1108@126.com

Y. Cao
School of Cyber Science and Engineering, Wuhan University, Wuhan, China
e-mail: yue.cao@whu.edu.cn

© The Author(s), under exclusive license to Springer Nature Singapore Pte Ltd. 2023
Y. Cao et al. (eds.), *Automated and Electric Vehicle: Design, Informatics and Sustainability*, Recent Advancements in Connected Autonomous Vehicle Technologies 3, https://doi.org/10.1007/978-981-19-5751-2_12

ICI	Intercarrier Interference
IoT	Internet Of Things
ISI	Intersymbol Interference
LDS	Low Density Signature
LDS-CDMA	Low Density Signature Code Division Multiple Access
LDS-OFDM	Low Density Signature Orthogonal Frequency Division Multiplex
MC-CDMA	Multi-Carrier Code Division Multiple Access
MPA	Message Passing Algorithm
MTC	Machine Type Communication
MUD	Multiuser Detection
MUI	Multiuser Interference
MUSA	Multiuser Shared Access
NOMA	Non-orthogonal Multiple Access
OFDM	Orthogonal Frequency Division multiplexing
OFDMA	Orthogonal Frequency Division Multiple Access
OMA	Orthogonal Multiple Access
PDMA	Pattern Division Multiple Access
PIC	Parallel Interference Cancellation
SCMA	Sparse Code Multiple Access
SIC	Successive Interference Cancellation
SISO	Soft-Input Soft-Output
TDMA	Time Division Multiple Access
VND	Variable Node Detector
V2V	Vehicle-To-Vehicle
K	Number of users
M	Data length of each user
N	Number of chips
Cn	The nth chip, also represents chip node
$d_{c;lds}$	Number of symbols that are superimposed at one chip
$d_{v;lds}$	Number of chips that are spread by one symbol
$v_{k;m}$	The mth symbol of the kth user, also represents variable node

12.1 What is NOMA

The explosive traffic growth in wireless communications has motivated many research activities in both academic and industrial communities. Following the large-scale commercialization of the forth generation (4G) networks, the fifth generation (5G) of mobile communications has become a focal point for global research and development [1, 2]. To satisfy requirements of future cellular systems, some enhanced technologies have been recently proposed for 5G, e.g. massive multiple-input multiple-output (MIMO), millimeter wave communications and ultra dense network. With resource demanding applications such as mobile internet and Internet of Things (IoT), air interface techniques need to be designed to achieve improved spectrum utilization and resource management [3]. Generally speaking, a new air interface for future cellular systems should consist of building blocks and configuration mechanisms such as advanced multiple access schemes, powerful forward error correction coding, adaptive multi-carrier waveforms and soon [4, 5]. With these block sand mechanisms,5G wireless networks can offer significant improvements in coverage and user experience, and are able to accommodate the a wide variety of user services, spectrum bands and traffic levels.

IoT, based on machine type communication (MTC), is considered as the next killer application for mobile networks. There will be an estimated at least 1000 billion Internet-capable devices by 2030. Although a business model has not yet established, measurements and monitoring for health, transportation, home appliances, smart grid, etc. are pushing the development of solutions for MTC and IoT. As a special type of MTC, vehicle-to-vehicle (V2V) communications have emerged with great interest due to the fact that they can help reduce traffic accidents and traffic jams. Some obvious benefits of V2V communication are enhanced road safety, increased commuter awareness of current traffic, weather, and road conditions, reduced delays at tollbooths, and the ability to enable groups of travelling cars to exchange multimedia information. For example, one vehicle can warn others about impending road hazards, adverse weather, or an upcoming traffic jam. The V2V wireless channel exhibits a very harsh signal propagation environment since both transmitter and receiver are in motion, and there are both mobile and stationary scatters as a result, this channel characteristic leads to a short channel coherence time. The dynamic nature of the V2V channel makes reliable communication a challenging task. Massive connectivity with a large number of devices is an important requirement for 5G networks. Sporadic traffics generated by MTC cause significant control signalling growth and network congestion. A machine usually disconnects from the network and returns to idle state immediately after data transmission. If it wants to connect to network and sends more data, it must initiate an entirely new connection, which greatly increases signalling traffic. This is also disadvantageous for delay sensitive MTC/V2V applications.

The goal in the design of cellular systems is to be able to accommodate as much traffic as possible (this is called capacity in cellular terminology) in a given bandwidth with some reliability. Multiple access technologies allow multiple sources communicate with the network simultaneously. In the history of wireless communications from the first generation (1G) to 4G, multiple access techniques have been the key to distinguish different wireless systems. It is well known that frequency division multiple access (FDMA) for the first generation (1G), time division multiple access (TDMA) mostly for the second generation (2G), code division multiple access (CDMA) for the third generation (3G), and orthogonal frequency division multiple access (OFDMA) for 4G are the primary multiple access techniques. In these conventional schemes, different users are allocated to orthogonal resources in either the time/frequency/code domain in order to avoid or alleviate inter user interference, thus they can be classified as orthogonal multiple access (OMA) techniques. In current mobile communication systems such as Long-Term Evolution (LTE) and LTE-Advanced, OMA techniques have been adopted, e.g. OFDMA. Ideally, no interference exists among multiple users due to the orthogonal resource allocation in OMA, simple detection techniques can thus be used to separate different users' signals. In other words, the users in each cell are allocated the resources exclusively and there is no inter-user interference, hence, low-complexity detection approaches can be implemented on the receiver side to retrieve the users' signals.

The fast growth of mobile internet has propelled more than 1000-fold data traffic increase for the cellular networks. Therefore, how to maximize the spectral efficiency becomes one of the key challenges to handle such explosive data traffic. Moreover, due to the rapid development of IoT, 5G systems need to support the massive connectivity of users and/or devices to meet the demand for low latency, low-cost devices, and diverse service types. Theoretically, it is known that OMA cannot always achieve the sum-rate capacity of multiuser wireless systems. Apart from that, in OMA schemes, the maximum number of supported users is limited by the degree of freedom and the scheduling granularity of orthogonal resources. This issue is more prominent when fairness among the users is considered [6]. For convenience, the system loading can be defined as follows [7, 8].

System loading. Consider a multiuser system with K users and N dimensions, where the dimensions mean any available degrees of freedom including chips, subcarriers, I/Q channels, antennas and polarisation. The system loading is described as the ratio of the number of supported users to the number of dimensions and is denoted as $\rho = K/N$. The system is respectively called in under-loaded, fully-loaded and overloaded conditions when $\rho < 1$, $\rho = 1$, and $\rho > 1$.

Recently, non-orthogonal multiple access (NOMA), including NOMA via power domain multiplexing and NOMA via code domain multiplexing, has been attracting a lot of attention. The main difference between these two groups of NOMA is whether utilizing the spreading technique. Different from conventional OMA, the NOMA schemes are highly expected to improve the spectral efficiency and accommodate much more users via non-orthogonal resource allocation. Basically, NOMA allows controllable interference by non-orthogonal resource allocation with a tolerable increase in the receiver complexity. This chapter aims to present a tutorial review of NOMA and compare them with OMA in various applications.

12.2 Advantages of NOMA

12.2.1 Improved Spectral Efficiency

In cellular systems, it is necessary to design efficient multiple access schemes that enable several multiple users to gain access and communicate simultaneously. There are a number of requirements that multiple access schemes must be able to meet, such as ability to handle several users without mutual interference and to maximize the spectrum efficiency. Figure 12.1 shows the throughput (capacity) comparison of

OMA and NOMA in downlink [9–12], where two users in the additive white Gaussian noise (AWGN) channel are considered as an example without loss of generality. The h_1 and h_2 are complex channel coefficients of the two users, the $N_{0,1}$ and $N_{0,2}$ are the power spectral density of Gaussian noise of the two users, the p_{total} is the sum of the two users' transmission power. It can be seen that the maximum total throughput is achieved when all the transmission power is allocated to user 1 only, which is achieved by both OMA and NOMA. However, the throughput region of NOMA is much wider than that of OMA. For example, if we want R_2 to be 0.8 b/s, the achievable R_1 for NOMA is approximately twofold higher than that for OMA. This is related to the fact that the throughput of user1 with a high $p_{\text{total}}|h_1|^2/N_{0,1}$ is bandwidth-limited rather than power-limited and superposition coding with user 2 allows user 1 to use the full bandwidth while being allocated only a small amount of transmission power because of power sharing with user 2. Thus, user 1 imparts only a small amount of interference to user 2. In contrast, OMA has to allocate a significant fraction of bandwidth to user 2 to increase its throughput, and this causes severe degradation in the throughput of user 1 whose throughput is bandwidth-limited. As for the uplink, Fig. 12.2 shows the throughput (capacity) comparison of OMA and NOMA with two users in the AWGN channel. We can see that the throughput region of NOMA is also wider than that for OMA. If we want R_2 to be 0.8 b/s, the achievable R_1 for NOMA is approximately 60% higher than that for OMA. Therefore, in both downlink and uplink transmissions, NOMA can significantly improve the spectral efficiency compared to OMA.

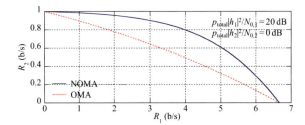

Fig. 12.1 Throughput (capacity) region in downlink

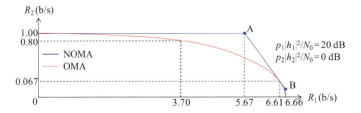

Fig. 12.2 Throughput (capacity) region in uplink

12.2.2 Massive Connectivity

In future cellular communications, the number of users or parallel data streams will inevitably exceeds the available dimensions as the demand for the spectrum is increasing while the bandwidth is fixed. Under such an overloaded condition, it is impossible to obtain the orthogonality of received signatures, consequently the performance of the bit error rate (BER), the block error rate (BLER) and the system throughput are limited by the severe multiuser interference (MUI). The non-orthogonal resource allocation in NOMA indicates that the number of supported users or parallel data streams is not strictly limited by the amount of available dimensions and their scheduling granularity [13, 14]. Therefore, in both downlink and uplink transmissions, NOMA can accommodate significantly more users than OMA by using non-orthogonal resource allocation [15–17]. In other words, NOMA can support fully-loaded and overloaded transmissions. For example, NOMA can achieve a reasonable good performance when the system loading is 200% [7, 13]. As for the maximum number of non-orthogonally multiplexed users, it is shown in [12] and [18] that when the system loading is 200%, the capacity gain of NOMA is significantly better than that of OMA, i.e., about 60% improvement compared with OMA. However, the further gain by continually increasing the system loading from 200 to 400% is relatively small, i.e., approximately 63% improvement compared with OMA. This indicates that it is sufficient to multiplex non-orthogonally a moderate number of users to obtain the most from the gain of NOMA.

12.3 Low Density Signature (LDS)

Wireless cellular technologies are continuously evolving to meet the increasing demands for the massive connectivity. For conventional multiple access techniques via code domain multiplexing, each user spreads the original data using a given spreading sequence, where elements of the spreading sequence usually take nonzero values, which are optimized under certain criteria, e.g., good auto-and/or cross-correlation properties. These spreading sequences are orthogonal to each other to avoid MUI, and naturally have high density, which means majority chips have nonzero values. The drawback is that, each user will see the interference coming from all other users at the chip level, and it cannot easily achieve satisfactory performance under overloaded conditions. To deal with these problems,a multiple access technique named low density signature has been proposed. Several milestones in this area have been achieved, leading to a flurry of further research.

12.3.1 LDS-CDMA

The LDS concept is first proposed for CDMA systems [19–24]. Figure 12.3 shows the block diagram of the LDS-CDMA transreceivers with K users transmitting to the same base station where the base station and each user are equipped with a single antenna. Let the processing gain of spreading to be N, and each user has a data vector consisting of M data symbols. We can see that after forward error correction (FEC) encoding and symbol mapping, the data are sent to a specially designed spreader. Instead of optimizing the N-chips sequences, the scheme intentionally arranges each user to spread its data, $v_{k,m}$, over very limited chips, c_n. More explicitly, the spreading sequences have small number of nonzero values and the rest chips are zero valued, hence the resultant signature matrix becomes very sparse. Basic principles of LDS are [25]:

- Changing the interference pattern being seen by each user.
- Limiting the amount of interference occurred on each user.

Figure 12.4 illustrates the LDS principle by using a simple exemplary system with 5 chips and 10 data symbols, where chip nodes and variable nodes respectively represent chips and data symbols. It can be seen that each symbol is spread over 2 chips. Each chip is used by 4 symbols that may belong to different users.

To elaborate the LDS structure more clearly, the sets of chip nodes and variable nodes are associated with chip node detector (CND) and variable node detector (VND), respectively. The iterative structure in the receiver is depicted in Fig. 12.5 and it closely follows the LDS in Fig. 12.4. CND and VND can be expressed by mathematical functions, and exchange soft messages through the edge interleave. It has been proved that LDS-CDMA significantly outperforms conventional CDMA systems under overloaded conditions [21].

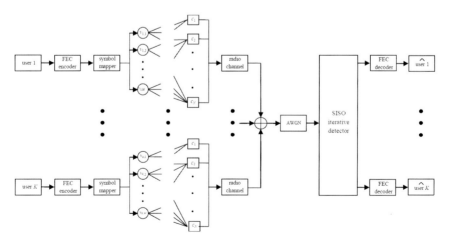

Fig. 12.3 LDS-CDMA system

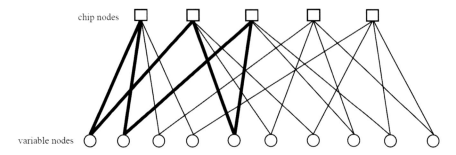

Fig. 12.4 Illustration of a LDS spreader

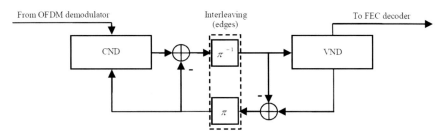

Key:

CND: Chip node detector (corresponding to the chip nodes)

VND: Variable node detector (corresponding to the user nodes)

Fig. 12.5 LDS iterative structure

12.3.2 LDS-OFDM

As an extension, LDS is applied to OFDM systems [18, 26–30]. Reference [31] extends LDS to a rateless scenario, i.e., the low density signature is a dynamic graph when data are transmitted. By doing so, the system spectral efficiency is improved, but the receiver complexity is higher than a fixed rate LDS. Figure 12.6 shows the block diagram of the LDS-OFDM transreceiver, where the functional blocks are similar to that of a conventional multicarrier code-division multiple-access (MC-CDMA) system. In MCCDMA, after FEC encoding and symbol mapping, each modulated symbol is multiplied with a spreading signature (a random sequence of chips) and subsequently OFDM modulation is arranged to modulate the chips onto respective subcarrier frequencies. Cyclic prefix (CP) has to be inserted to eliminate ISI and ICI. Compared with conventional MC-CDMA transmitter, the main difference in Fig. 12.6 is that the spreading signature has low density by the use of zero padding, which means a large number of chips in the sequence are zeros. Due to the LDS structure, each data symbol is spread over very limited chips. Each user's generated chip is transmitted over an orthogonal subcarrier, and each subcarrier is only used by a limited number of data symbols that may belong to different users. Each user,

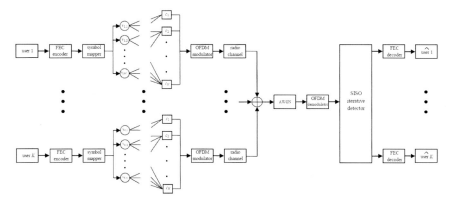

Fig. 12.6 LDS-OFDM system

transmitting on given subcarriers, will experience interference from only a small number of other users' data symbols. More explicitly, the number of symbols that are superimposed on each chip is much less than the total number of symbols, $d_{c,lds} \ll (K \times M)$, where $d_{c,lds}$ is the number of symbols that are superimposed at one chip. Meanwhile, the number of chips that are spread by each symbol is much less than the total number of chips, $d_{v,lds} \ll N$, where $d_{v,lds}$ is the number of chips that are spread by one symbol. In other words, the spreading sequences have a maximum of $d_{v,lds}$ nonzero values and $N - d_{v,lds}$ zero values, then they are interleaved uniquely for each user such that the resultant signature matrix becomes sparse. The interleaving process is designed so that at each received chip there exists a contribution of, instead of all users, only a small number of users. Consequently, the interference pattern being seen by each user is different [32].

It can be seen that the philosophy of LDS is that the number of users symbols which are superimposed on each chip is much less than the total number of data symbols. Meanwhile, the number of chips that are spread by each symbol is much less than the total number of chips. As a result, if a fraction of signal of some user is superimposed by a fraction of signals coming from a relatively small number of interferers, then the search-space should be moderate, thus detection technique with affordable complexity can be used to recover the corrupted signal. Moreover, apart from being practical for implementation, the LDS structure also benefits from having the intrinsic interference diversity by avoiding strong interferers to corrupt all chips of a user. Therefore, LDS is an effective technique for fully-loaded and overloaded transmissions [33–35]. The drawback of LDS is that its performance degrades with high order constellations.

12.4 Sparse Code Multiple Access (SCMA)

Although LDS performs well under overloaded conditions, its performance is not ideal with high-order constellations. Based on the LDS technique, SCMA is proposed as an enhanced version of LDS-OFDM [36–38].

12.4.1 System Model

In LDS-OFDM, a LDS spreader expands a QAM symbol to a sequence of complex symbols by using a given low density signature. Hence, a LDS spreader can be seen as a process in which a number of coded bits are mapped to a sequence of complex symbols. From this point of view, the QAM mapper block and the LDS spreader can be merged together to directly map a set of bits to a complex vector so called a SCMA codeword. In other words, according to predefined SCMA codebook sets, bit streams are directly mapped to sparse codewords, and the whole process can be interpreted as a coding procedure from the binary domain to a multidimensional complex domain as shown in Fig. 12.7. With this interpretation, a simple LDS spreading action is generalized to a coding process which in turn raises a new problem in terms of complex multidimensional codeword design rather than a relatively simple low density signature design [39–41]. The SCMA characters can be summarized as follows:

- Binary domain data are directly encoded to multidimensional complex domain codeword selected from a predefined codebook set.
- Multiple access is achievable by generating multiple codebooks, one for each layer or user.
- Codeword of the codebook is sparse such that the message passing algorithm (MPA) multiuser detection technique is applicable to detect the multiplexed codeword with a moderate complexity.
- Like LDS-OFDM, SCMA can be overloaded such that the number of multiplexed layers can be more than the spreading factor.

Figure 12.8 shows a SCMA with 6 users where each user has a predefined codebook. All codewords in the same codebook contain zeros in the same two dimensions, and the positions of zeros in different codebooks are distinct to facilitate the collision avoidance of any two users. For each user, two bits are mapped to a complex codeword. Codewords for all users are multiplexed over four shared orthogonal resources (e.g., OFDM subcarriers).

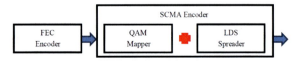

Fig. 12.7 Merging of QAM modulator and LDS spreading in a SCMA encoder

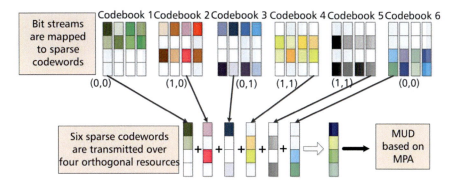

Fig. 12.8 SCMA system

The main difference between LDS-OFDM and SCMA is that a multi-dimensional constellation is designed for SCMA to generate codebooks, which brings the shaping gain that is not possible for LDS-OFDM [42]. Here, shaping gain is the gain in the average symbol energy when the shape of a constellation is changed. For the concatenated approach with high-order constellations, the multidimensional constellation can be optimized to obtain shaping gain, then codebooks are generated based on the multi-dimensional constellation. Multi-dimensional lattice constellation design is a challenging problem which has been studied in different aspects of communications. The SCMA codebook design is even more complicated as multiple layers are multiplexed with different codebooks. A SCMA encoder is defined as $v = f(b)$, where b and v respectively represent the bit sequence and the SCMA codeword vector. The N dimensional complex codeword v is a sparse vector with $d_{v,lds} \ll N$ non-zero entries. Let c denote an $d_{v,lds}$ dimensional complex constellation point such that: $c = g(b)$. A SCMA encoder can be redefined as $f \triangleq sg$ where s simply maps the $d_{v,lds}$ dimensions of a constellation point to an N dimensional SCMA codeword. In fact, s is the low density spreading in LDS. The resulting codebook contains of M codewords each consisting of N complex values from which only maximum $d_{v,lds}$ are non-zero specified by s. The difference between LDS and SCMA is that LDS only consider spreading, i.e., $f \triangleq s$, while SCMA considers both spreading and constellation such that $f \triangleq sg$. The design criteria of multi-dimensional constellations include:

- Minimization of the average energy per constellation point.
- Maximization of the diversity order.
- Maximization of the minimum product distance.
- Minimization of the product kissing number for the product distance.

As the appropriate design criterion and specific solution to the multi-dimensional problem are still unknown, a multi-stage approach has been proposed to realize a suboptimal solution. Specifically, a complex constellation which is called the mother constellation is first optimized to improve the shaping gain, and then some codebook-specific operations are performed to the mother constellation to generate the constellation for each codebook. Three typical operations are phase rotation, complex conjugate, and dimensional shuffling of the constellation. In the generated constellations after codebook-specific operations, each constellation point is multiplied with a low density matrix to generate a codeword. In this way, SCMA codebooks can be obtained.

On receiver side, the SCMA detection is the same to that of the LDS detection [27]. The algorithm iteratively updates the belief associated to the edges in the factor graph by passing the extrinsic information of constellation points between resource nodes and layer nodes. Note that resource nodes and layer nodes in SCMA respectively correspond to chip nodes and variable nodes in LDS. At each resource node of SCMA, the received signal is combined with the extrinsic information passed by the resource nodes through the rest of the edges connected to that resource node. At each layer node, the a priori information of the transmitted layers (usually assumed to be uniformly distributed) is combined with the extrinsic information passed by the resource nodes through the rest of edges connected to that layer node. Finally, after a few iterations, the detector converges to reliable soft information of the transmitted layers [43–45].

There are works on SCMA such as the blind detection for uplink grant-free multiple access [46] and irregular SCMA schemes [47]. SCMA is also applied in physical layer security transmissions [48]. Although SCMA performs better than LDS, the codebooks still need to be improved by the means of graph optimization and the constellation design.

12.4.2 Multi-Stage Optimization Approach of Codebook Design

The codebook design problem of SCMA can be defined as

$$\Gamma = \arg\max f\left([\mathbf{S}_k]_{k=1}^K, [\mathbf{g}_k]_{k=1}^K; K, M, d_{v,lds}, N\right) \tag{12.1}$$

Reference [42] points out that the solution of this multi-dimensional problem is still unknown, a multi-stage optimization approach is the suboptimal solution for the issue. We summarize the multi-stage optimization procedures as follows.

(1) *Low Density Signature*: As described before, the mapping matrix is actually the low density signature of the LDS scheme, and it is used to determine the number of layers interfering at each resource node which in turn defines the complexity of the MPA detection. The sparser the codewords structure, the lower complexity is the MPA detection. Optimization of the mapping matrix is the same as the LDS scheme.

(2) *Constellation Points*: Having the mapping matrix $\mathbf{S}^+$, the optimization problem of an SCMA is reduced to

$$\mathbf{g}^+ \arg\max f\left(\mathbf{S}^+, \mathbf{g}; K, M, d_{v,lds}, N\right) \tag{12.2}$$

12 The Overview of Non-orthogonal Multiple Access … 231

The problem is to define K different $d_{v,lds}$ dimensional constellations each containing M points. To simplify the optimization problem, the constellation points of the layers are modeled based on a mother constellation and the layer-specific operators, i.e. $g_k \triangleq (\Delta_k)g$, $k \in [1, K]$, where Δ_k denotes a constellation operator. According to the model, the SCMA code optimization problem turns into

$$g^+, \left[\Delta_k^+\right]_{k=1}^K = \arg\max f\left(\mathbf{S}^+ g = \left[(\Delta_k^+)g\right]_{k=1}^K; K, M, d_{v,lds}, N\right) \qquad (12.3)$$

As a suboptimal approach to the above problem, the mother constellation and the operators are determined separately.

(3) *Mother Multi-dimensional Constellation*: The target is to design a multi-dimensional lattice constellation. A compact multi-dimensional constellation can be designed by minimizing the average alphabet energy for a given minimum Euclidian distance between the constellation points. A unitary operation (lattice rotation) can be applied directly on the lattice constellation without sacrificing the Euclidian distance of the constellation. The lattice rotation can be optimized to improve the product distance of the constellation and induce dependency among the lattice dimensions, and reduce the number of projections per lattice dimensions to reduce the detection complexity. Also, the dimensional power properties of the constellation can be changed by lattice rotation, for better convergence of the MPA receiver by taking advantage of near-far effect of the overlaid codeword. Shaping gain of multi-dimensional constellations is the major source of the performance gain of SCMA over LDS. After optimizing the constellation set, the corresponding constellation function is defined to setup the mapping rule between the binary words and the constellation alphabet points. Following the Gray mapping rule, the binary words of any two closet constellation points can have a Hamming distance of 1. The procedures of mother multi-dimensional constellation include following steps:

(i) Design metrics and rotated constellations

Large minimum Euclidean distance of a multi-dimensional constellation ensures a good performance of the SCMA system with a small number of layers where there are no collisions between the layers over a resource node. Once the number of layers grows, two or more layers may collide over a resource node. Under this condition, it is important to induce dependency among the nonzero elements of codewords to be able to recover colliding codewords from the other resource nodes. In addition, power imbalance across the dimensions of codewords introduces near-far effect among colliding layers. It helps MPA detector to operate more effectively to remove interferences among layers.

Having a constellation with a desirable Euclidian distance profile, a unitary rotation can be applied to the base constellation in order to control dimensional dependency and power variation of the constellation while maintaining the Euclidian distance profile. Similar to the code design for communications over fading channels, a unitary rotation can be applied to maximize the minimum product distance of the constellation. Therefore, the design objective encapsulates both the sum distance and the product distance between the points in the mother constellation.

(ii) Rotated lattice constellations

In general, the base constellation can be any multi-dimensional constellation with a maximized minimum Euclidean distance. At low rates, constellation can be designed by heuristic optimization. However, at medium and higher rates, a structured construction way is required. In fact, lattice constellation is a structural approach of the base constellation design. As a special case of lattice constellations, we can consider the base constellation to be formed by orthogonal QAMs on different complex planes. It is equivalent to a constellation from the lattice. Gray labeling is an advantage of this type of lattice constellations.

(iii) Shuffling multi-dimensional constellations in real and imaginary axes

If a complex constellation is built such that its real part is independent of its imaginary part, it can help reduce the decoding complexity while maintaining dependency among the complex dimensions of the resulted multi-dimensional constellation. Using this technique, the complexity order of MPA reduces from $M^{dc,lds}$ to $M^{dc,lds/2}$ which results in significant complexity saving especially for large constellation sizes.

A shuffling is to construct a $d_{v,lds}$ dimensional complex mother constellation from Cartesian product of two $d_{v,lds}$ dimensional real constellations, where each of them is constructed by the same method described in the previous section. One of these two $d_{v,lds}$ dimensional real constellations corresponds to the real part of the points of the complex mother constellation and the other one corresponds to the imaginary part.

Figure 12.9 shows an example of the shuffling to construct a 16-point SCMA mother constellation with two nonzero positions ($d_{v,lds} = 2$). The optimum rotation angle that maximizes the minimum product distance is $\tan^{-1}((1 + \sqrt{5})/2)$ [42]. First, two QPSK constellations are rotated using the optimum angle. Each point of the rotated constellations can be queued according to the axis value: X1(11 01 10 00), X2(10 11 00 01), Y1(11 01 10 00), Y2(10 11 00 01). Subsequently, a shuffling is performed to separate real and imaginary parts. One of these two real constellations corresponds to the real part of the points of the complex mother constellation and the other one corresponds to the imaginary part.

To explain the SCMA principle more clearly, we show a 150% loaded SCMA as an example. There are 6 users and 4 subcarriers, each user is spread on 2 subcarriers, thus $N = 4$, $K = 6$ and $d_{v,lds} = 2$. We use the 16-point constellation presented in Fig. 12.9 as the constellation scheme. The LDS matrix is shown in (12.4), where $a_0 = 1$, $a_1 = \exp(j2\pi/27)$, $a_2 = \exp(j4\pi/27)$ [20]. It should be noted that each row and each column of the LDS matrix in (12.4) represent a subcarrier/chip and a user, respectively. Table 12.1 illustrates the generated SCMA codewords which is a combination of the LDS matrix in (12.4) and the constellation in Fig. 12.9.

Fig. 12.9 16-point SCMA constellation

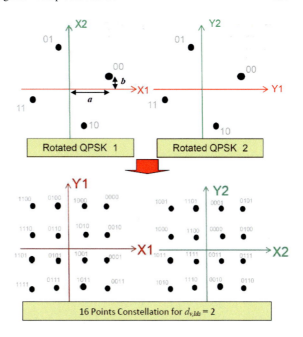

$$\begin{bmatrix} a_0 & a_1 & a_2 & 0 & 0 & 0 \\ a_1 & 0 & 0 & a_0 & a_2 & 0 \\ 0 & a_2 & 0 & a_1 & 0 & a_0 \\ 0 & 0 & a_0 & 0 & a_1 & a_2 \end{bmatrix} \tag{12.4}$$

(iv) *Rotation to Minimize Number of Projection Points*

For the sake of simplicity of the MPA detection, it is more desirable to use mother constellations that have a smaller number of projections per resource node (or complex dimension). Let m denote the number of projections per complex dimensions of an M point constellation. It is obvious that $m < M$. As m decreases, the complexity of the corresponding MPA detector is also reduced by $m^{dc,lds}$. During the process of mother constellation design, the rotation matrix can be set in a way so as to lead to

Table 12.1 SCMA Codewords for 16-point

	Information bits	SCMA codeword
User 1	0000	$[(a + jb)a_0, (b + jb)a_1, 0, 0]$
User 2	0101	$[(-b - jb)a_1, 0, (a + ja)a_2, 0]$
User 3	0011	$[(a - ja)a_2, 0, 0, (b - jb)a_0]$
User 4	0101	$[0, (-b - jb)a_0, (a + ja)a_1, 0]$
User 5	1100	$[0, (-a + ja)a_2, 0, (-b + jb)a_1]$
User 6	0111	$[0, 0, (-b - ja)a_0, (a - jb)a_2]$

the lower number of projected points. It makes the minimum product distance equal to zero and degrades the performance of the SCMA system. Consequently, there is a trade-off between the performance requirement and the complexity in this case. As an example, Fig. 12.10 shows a solution with 9-projections per complex dimension of a 16-point constellation [42].

Similarly, we show the generated SCMA codewords for 16-points with 9-projection-point in Table 12.2. Such SCMA codebook is a combination of the LDS matrix and the constellation which are respectively presented in (12.4) and Fig. 12.10.

(4) *Constellation Function Operators*: By having a solution for the mother constellation, the original SCMA optimization problem is further reduced to

$$\left[\Delta_k^+\right]_{k=1}^K = \arg\max f\left(\mathbf{S}^+, \mathbf{g} = \left[(\Delta_k)\mathbf{g}^+\right]_{k=1}^K; K, M, d_{v,lds}, N\right) \quad (12.5)$$

Here, we limit the operators to those with unitary representation over real domain which guarantees that the Euclidian distances between different codewords are not

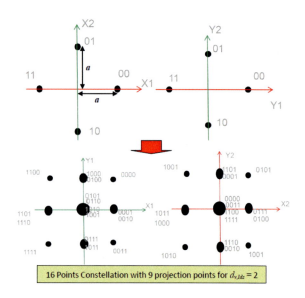

Fig. 12.10 16-point SCMA constellation with 9-projection-point

Table 12.2 SCMA Codewords for 16-point with 9-projection-point

	Information bits	SCMA codeword
User 1	0000	$[(a + ja)a_0, (0 + j0)a_1, 0, 0]$
User 2	0101	$[(0 + j0)a_1, 0, (a + ja)a_2, 0]$
User 3	0011	$[(a - ja)a_2, 0, 0, (0 + j0)a_0]$
User 4	0101	$[0, (0 + j0)a_0, (a + ja)a_1, 0]$
User 5	1100	$[0, (-a + ja)a_2, 0, (0 + j0)a_1]$
User 6	0111	$[0, 0, (0 - ja)a_0, (a + j0)a_2]$

altered. Three typical operators are complex conjugate, phase rotation, and dimensional permutation of the lattice constellation. The codebooks of different SCMA layers are constructed based on the mother constellation g and a layer-specific operator Δ_k for layer k. The task of the MPA detector is to separate the interfering symbols in an iterative fashion. As a basic rule, interfering symbols at a resource node are more easily separated if their power level is more diverse. Intuitively, the strongest symbol is first detected and then it helps the rest to be detected by removing the next strongest symbols. Based on this reasoning, the mother constellation must have a diverse average power level over the constellation dimensions. This target can be achieved by appropriate rotation of the lattice constellation. Assuming the dimensional power level of the mother constellation is diverse enough, the permutation operators of the SCMA codebooks must be selected in a way to capture as much power diversity as possible over the interfering layers. The power variation over the layers can be optimized following the approach described in the following: the permutation of each codebook set is designed to avoid interfering the same dimensions of a mother constellation over a resource node.

12.5 Multiuser Shared Access (MUSA)

MUSA is another NOMA technique via code domain multiplexing, and it is suitable for overloaded transmissions [49]. Figure 12.11 shows the MUSA system. Multiple spreading sequences constitute a pool from which each user can randomly pick one of the sequences. Note that for the same user, different spreading sequences may also be used for different symbols, which may further improve the performance via interference averaging. Then all spreading symbols are transmitted over the same time–frequency resources. The spreading sequences should have low cross-correlation and can be M-ary. MUSA differs from MC-CDMA in that it is basically synchronous transmission mechanism when users signals arrive at the base station, while MC-CDMA doesn't have this kind of synchronism requirement in the uplink. In addition, MUSA uses non-binary spreading sequences, while binary spreading sequences are usually considered in classical MC-CDMA systems.

In downlink MUSA, users are separated into different groups. In each group, different users' symbols are mapped to different constellations in a way to ensure Gray mapping in the combined constellation of superposed signals. The combined constellation is determined not only by the modulation order of each user, but also by the transmit power partition among multiplexed users. Orthogonal sequences can be used to spread the superposed symbols to obtain time or frequency diversity gain. Gray mapping of the combined constellation reduces the reliance on advanced receivers, therefore less processing-intensive receivers such as symbol-level SIC can be used.

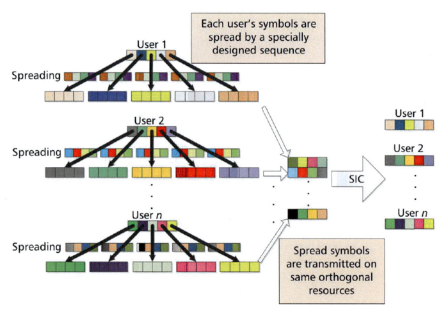

Fig. 12.11 MUSA system

12.6 Pattern Division Multiple Access (PDMA)

PDMA is a NOMA scheme that can be realized in multiple domains [50]. At the transmitter, PDMA uses non-orthogonal patterns which are designed to maximize the diversity and minimize the overlaps of multiple users. Then, multiplexing can be realized in code domain, power domain, space domain or their combinations. Multiplexing in code domain is similar to that in LDS, but the number of subcarriers connected to the same symbols in the factor graph can be different. At the receiver, MPA is performed for interference cancelation. In the case of multiplexing in power domain, power allocation needs to be carefully considered under the total power constraint. SIC can be also used at the receiver according to SNR difference among multiplexed users. Multiplexing in space domain, i.e., spatial PDMA, can be combined with the multiple-antenna technique. The advantage of spatial PDMA compared with multiuser MIMO is that PDMA doesn't require joint precoding to realize spatial orthogonality, which significantly reduces system complexity. In addition, multiple domains can be combined in PDMA to make full use of various available wireless resources. MPA is used at the receiver to separate different users' signals. The design principle for spreading matrices in code domain PDMA is to facilitate interference cancelation. The system model of code domain PDMA is similar to that of MUSA,but the spreading matrix in MUSA is designed with the following principles:

12 The Overview of Non-orthogonal Multiple Access …

- The number of groups with different number of nonzero elements in the spreading sequence should be maximized.
- The number of the overlapped spreading sequences which have the same number of nonzero elements should be minimized.

12.7 Comparison

In this chapter, we have reviewed different code domain NOMA schemes, including LDS-CDMA, LDS-OFDM, SCMA, MUSA and PDMA. Advantages and drawbacks of these multiple access techniques are discussed. Table 12.3 summarizes these techniques and their comparisons.

The research activities of code domain NOMA have attracted increasing attention due to their ability to support overloaded transmissions. With the contributions in mind, there are several open issues needed to be figured out.

- Although code domain NOMA including LDS and SCMA have been analyzed and optimized in many papers, the codebook design and optimization are still open issues. Different from techniques of sparse graph coding, LDS and SCMA are multiple access schemes for the multiuser scenario, and the channel condition of different users is random and dynamic. Thus the research in this area is more challenging than conventional sparse graphs. To date, there does not exist a systematic approach for the system design. It is thus of vital importance to establish a unified framework for the design and optimization of NOMA schemes.
- Due to the use of MPA for the detection, the receiver complexity is a big challenge in practical applications, especially when the size of the sparse graph is large or the constellation order is high. It is necessary to continue the research on receiver design, by finding approaches other than MPA to reduce the receiver complexity while maintaining satisfactory performance.
- The existing code domain NOMA schemes are mainly applied in OFDM systems. In future cellular networks, more advanced waveforms may be utilized. Hence, it is urgent to research the combination of NOMA techniques and advanced waveforms other than OFDM.

Table 12.3 Comparisons of code domain NOMA techniques

	LDS-CDMA	LDS-OFDM	SCMA	MUSA	PDMA
Multi-carrier	No	Yes	Yes	Yes	Yes
Spreading	Yes	Yes	Yes	Yes	Yes
Non-binary chips	No	No	No	Yes	No
Constellation shaping	No	No	Yes	No	No
Sparse structure	Yes	Yes	Yes	Yes	Yes
Handle overloading	Yes	Yes	Yes	Yes	Yes
Receiver	MPA	MPA	MPA	SIC	MPA

- SCMA blind detection should be further developed to include decoding of user's data with no complete knowledge of SCMA codebook sets. This is of particular interest to the vehicular communications, where the users' number is dynamic, and the roadside access point is random. This necessitates the blind detection of LDS sparse graph or SCMA codebooks, the MPA cannot be performed otherwise.

References

1. J. Gozalvez, Prestandard 5G developments [Mobile Radio]. IEEE Veh. Technol. Mag. **9**(4), 14–28 (2014)
2. G. Jiann-Ching, L. Pei-Kai, C. Yih-Shen, A. Hsu, H. Chien-Hwa, L. Gabriel, On 5G radio access architecture and technology [Industry Perspectives]. IEEE Wirel. Commun. **22**(5), 2–5 (2015)
3. A. Osseiran, F. Boccardi, V. Braun, K. Kusume, M. Patrick, M. Michal, Scenarios for 5G mobile and wireless communications: the vision of the METIS project. IEEE Commun. Mag. **52**(5), 26–35 (2014)
4. C.W. Hau, F. Zhong, R. Haines, Emerging technologies and research challenges for 5G wireless networks. IEEE Signal Process. Lett. **21**(2), 106–112 (2014)
5. F. Boccardi, R.W. Heath, A. Lozano, T.L. Marzetta, P. Petar, Five disruptive technology directions for 5G. IEEE Commun. Mag. **52**(2), 74–80 (2014)
6. G. Wunder, P. Jung, M. Kasparick, T. Wild, 5GNOW: non-orthogonal, asynchronous waveforms for future mobile applications. IEEE Commun. Mag. **52**(2), 97–105 (2014)
7. M. Al-Imari, P. Xiao, M. Imran, Uplink non-orthogonal multiple access for 5G wireless networks. *International Symposium on Wireless Communications Systems*, pp. 781–785(2014)
8. L. Wen. Non-orthogonal multiple access schemes for future cellular systems. Ph.D. dissertation, University of Surrey (2016)
9. Y. Saito, Y. Kishiyama, A. Benjebbour, T. Nakamura, A. Li, H. Kenichi, Non-orthogonal multiple access (NOMA) for cellular future radio access. *IEEE Vehicular Technology Conference* (2013)
10. X. Chen, A. Beijebbour, A. Li, H. Jiang, H. Kayama, Consideration on successive interference canceller (SIC) receiver at cell-edge users for non-orthogonal multiple access (NOMA) with SU-MIMO. *IEEE International Symposium on Personal, Indoor, and Mobile Radio Communications* (2015)
11. A. Benjebbour, A. Li, K. Saito, Y. Saito, Y. Kishiyama, T. Nakamura, NOMA: from concept to standardization. *IEEE Conference on Standards for Communications and Networking* (2015)
12. K. Higuchi, A. Benjebbour, Non-orthogonal multiple access (NOMA) with successive interference cancellation for future radio access. IEICE Trans. Commun. **98**(3), 403–414 (2015)
13. D. Linglong, W. Bichai, Y. Yifei, H. Shuangfeng, I. Chih-Lin, W. Zhaocheng, Non-orthogonal multiple access for 5G: solutions, challenges, opportunities, and future research trends. IEEE Commun. Mag. **53**(9), 100–103 (2015)
14. S. Timotheou, I. Krikidis, Fairness for non-orthogonal multiple access in 5G systems. IEEE Signal Process. Lett. **22**(10), 1647–1651 (2015)
15. S. Qi, H. Shuangfeng, I. Chin-Lin, P. Zhengang. Energy efficiency optimization for fading MIMO non-orthogonal multiple access systems. *IEEE International Conference on Communications* (2015)
16. X. Xiong, W. Xiang, K. Zheng, H. Shen, X. Wei, An open source SDR-based NOMA system for 5G networks. IEEE Wirel. Commun. **22**(6), 24–32 (2015)
17. D. Zhiguo, P. Mugen, H.V. Poor, Cooperative non-orthogonal multiple access in 5G systems. IEEE Commun. Lett. **19**(8), 1462–1465 (2015)

18. R. Razavi, M. Al-Imari, M.A. Imran, R. Hoshyar, C. Dageng, On receiver design for uplink low density signature OFDM (LDS-OFDM). IEEE Trans. Commun. **60**(11), 3499–3508 (2012)
19. A. Montanari, D. Tse, Analysis of belief propagation for non-linear problems: The example of CDMA, or: how to prove tanaka's formula. *IEEE Information Theory Workshop* (2006)
20. J. van de Beek, B.M. Popovic, Multiple access with low-density signatures. *IEEE Global Telecommunications Conference* (2009)
21. R. Hoshyar, F. Wathan, R. Tafazolli, Novel low-density signature for synchronous CDMA systems over AWGN channel. IEEE Trans. Signal Process. **56**(4), 1616–1626 (2008)
22. F. Wathan, R. Hoshyar, R. Tafazolli. Iterated SISO MUD for synchronous un-coded CDMA systems over AWGN channel: performance evaluation in overloaded condition. *International Symposium on Communications and Information Technologies*, pp. 397–402(2007)
23. R. Hoshyar, F. Wathan, R. Tafazolli, Dynamic grouped chip-level iterated multiuser detection based on Gaussian forcing technique. IEEE Commun. Lett. **12**(3), 167–169 (2008)
24. L. Wen, M. Su, Joint sparse graph over GF(q) for code division multiple access systems. IET Commun. **9**(5), 707–718 (2015)
25. L. Wen, J. Lei, M. Su, Improved algorithm for joint detection and decoding on the joint sparse graph for CDMA systems. IET Commun. **10**(3), 336–345 (2016)
26. J. Choi, Low density spreading for multicarrier systems. *IEEE Eighth International Symposium on Spread Spectrum Techniques and Applications*, pp. 575–578(2004)
27. R. Hoshyar, R. Razavi, M. Al-Imari, LDS-OFDM an efficient multiple access technique. *IEEE 71st Vehicular Technology Conference* (2010)
28. M. Al-Imari, M.A. Imran, R. Tafazolli, C. Dageng, Subcarrier and power allocation for LDS-OFDM system. *IEEEVehicular Technology Conference* (2011)
29. L. Wen, R. Razavi, M.A. Imran, P. Xiao, Design of joint sparse graph for OFDM system. IEEE Trans. Wireless Commun. **14**(4), 1823–1836 (2015)
30. L. Wen, R. Razavi, P. Xiao, M.A. Imran, Fast convergence and reduced complexity receiver design for LDS-OFDM system. *IEEE International Symposium on Personal, Indoor and Mobile Radio Communicaitons* (2014)
31. Z. Zhaoyang, C. Shaolei, W. Kedi, Non-coded rateless multiple access. *International Conference on Wireless Communications & Signal Processing* (2012)
32. L. Wen, T. Wang, J. Lei, Receiver optimization on non-binary joint sparse graph for OFDM system. Wirel. Commun. Mob. Comput. **16**(3), 3360–3376 (2016)
33. L. Wen, R. Razavi, M.A. Imran, P. Xiao, Joint sparse graph for FBMC/OQAM systems. IEEE Transactions on Vehiclular Technology **67**(7), 6098–6112 (2018)
34. L. Wen, R. Razavi, J. Lei, K. Lai, J. Zhong, Uplink multi-carrier multiple access scheme LDS-IOTA. IET Commun. **13**(14), 2163–2167 (2018)
35. L. Wen, R. Razavi, J. Lei, Intrinsic interference use for FBMC-IOTA systems. Applied Science **9**(3), 23–35 (2019)
36. H. Nikopour, H. Baligh, Sparse code multiple access. *IEEE International Symposium on Personal Indoor and Mobile Radio Communications* (2013)
37. H. Nikopour, E. Yi, A. Bayesteh, K. Au, H. Mark, B. Hadi, M. Jianglei, SCMA for downlink multiple access of 5G wireless networks. *IEEE Global Communications Conference* (2014)
38. U. Vilaipornsawai, Z. Liqing, H. Nikopour, E. Yi, A. Bayesteh, Uplink contention based SCMA for 5G radio access. *IEEE Global Communications Conference Workshops* (2014)
39. K. Lai, L. Wen, J. Lei, G. Chen, P. Xiao, A. Maaref, Codeword position index based sparse code multiple access system. IEEE Wirel. Commun. Lett. **8**(3), 737–740 (2019)
40. K. Lai, L. Wen, Secure transmission with interleaver for uplink sparse code multiple access system. IEEE Wirel. Commun. Lett. **8**(2), 336–339 (2019)
41. K. Lai, L. Wen, J. Lei, J. Zhong, G. Chen, X. Zhou, Simplified sparse code multiple access receiver by using truncated messages. IET Commun. **12**(16), 1937–1945 (2018)
42. M.Taherzadeh, H. Nikopour, A. Bayesteh, H. Baligh, SCMA codebook design. *IEEE Vehicular Technology Conference* (2014)
43. K. Lai, L. Wen, J. Lei, P. Xiao, A. Maaref, M.A. Imran, Sub-graph based joint sparse graph for sparse code multiple access systems. IEEE Access **6**(5), 25066–25080 (2018)

44. K. Lai, J. Lei, L. Wen, G. Chen, Codeword position index modulation design for sparse code multiple access system. IEEE Trans. Veh. Technol. **69**(11), 13273–13288 (2020)
45. T. Wang, L. Wen, Y. Zhou, A energy balanced routing scheme in wireless sensor networks based on non-uniform clustering. Sensor Networks **27**(4), 239–249 (2018)
46. A. Bayesteh, E. Yi, H. Nikopour, H. Baligh, Blind detection of SCMA for uplink grant-free multiple-access. *International Symposium on Wireless Communications Systems* (2014)
47. Z. Shutian, X. Baicen, X. Kexin, C. Zhiyong, X. Bin, Design and analysis of irregular sparse code multiple access. *International Conference onWireless Communications & Signal Processing* (2015)
48. K. Lai, J. Lei, L. Wen, G. Chen, W. Li, P. Xiao, Secure transmission with randomized constellation rotation for downlink sparse code multiple access system. IEEE Access **6**(99), 5049–5063 (2017)
49. Z. Yuan, G. Yu, W. Li, Multi-user shared access for 5G. Telecommun. Network Technol. **5**(5), 28–30 (2015)
50. Z. Jie, L. Bing, S. Xin, R. Liping, X. Rongrong, Pattern division multiple access (PDMA) for cellular future radio access. *International Conference on Wireless Communications & Signal Processing* (2015)

Chapter 13
Decentralized Trust Management System for VANETs

Yu Wang, Jianyong Song, Yu'ang Zhang, and Yue Cao

Abstract Due to the high mobility and variability of Vehicular Ad Hoc Networks (VANETs), heterogeneous vehicles are usually strangers and cannot fully trust each other. This problem becomes more serious when there are malicious vehicles in the VANETs. Therefore, how to effectively evaluate the reliability of vehicles is an important issue in the Internet of vehicles. The key step is to establish an effective trust mechanism, judge whether the message sender is trustworthy and check whether the information content is trustworthy before taking any measures, and spread accurate, real, and up-to-date trust content among network entities. This chapter first introduces the basic concepts of trust in VANETs, then reviews analyze, and compares some recently proposed trust management mechanisms. Finally, this chapter discusses the current challenges and future research hotspots of VANETs trust management.

Keywords Security · Trust management · Vehicular ad-hoc networks · Malicious attacks

Abbreviations

ITS   Intelligent Transportation System
VANETs   Vehicular Ad hoc Networks
RSU   Road Side Unit

Y. Wang (✉) · J. Song · Y. Zhang · Y. Cao
School of Cyber Science and Engineering, Wuhan University, Wuhan, China
e-mail: wang.yu@whu.edu.cn

J. Song
e-mail: zaiyu404@qq.com

Y. Zhang
e-mail: yuang.zhang@whu.edu.cn

Y. Cao
e-mail: yue.cao@whu.edu.cn

© The Author(s), under exclusive license to Springer Nature Singapore Pte Ltd. 2023
Y. Cao et al. (eds.), *Automated and Electric Vehicle: Design, Informatics and Sustainability*, Recent Advancements in Connected Autonomous
Vehicle Technologies 3, https://doi.org/10.1007/978-981-19-5751-2_13

OBU	On board Unit
TA	Trusted Authority
DoS	Denial of Service
DDoS	Distributed Denial of Service
DTMS	Distributed Trust Management System

13.1 Introduction

With the rapid development of science and technology, Intelligent Transportation System (ITS) has been emerging cutting use cases for future mobility.

When it comes to the intelligent transportation system, intelligent vehicles and VANETs are indispensable. VANETs have the potential to change the way people travel by creating a secure wireless communication network that includes cars, buses, traffic lights, mobile phones, and other devices [1]. VANETs can not only provide entertainment services on vehicles but also improve road safety and traffic efficiency (for example, real-time sharing of road traffic conditions, when there is a traffic accident in front, the front vehicle will timely transmit the information to the rear vehicle, to avoid greater traffic accidents).

However, due to the increasing dependence on wireless communication, and VANETs have the characteristics of high mobility and network variability, they are vulnerable to security threats for VANETs. The security and privacy challenges of VANETs include integrity, confidentiality, non-repudiation, authenticity, availability, and privacy protection. The traditional encryption algorithm is usually used to solve part of the security problems (e.g., privacy protection), but the trust problems (e.g., identify malicious vehicles which possess valid certificates) can not be solved well [2].

The trust management system usually calculates the trust value of vehicles and messages through certain methods. If the trust value of the vehicle is greater than the threshold, the vehicle is trustworthy, otherwise, it is untrustworthy. The existing trust management system can usually be divided into entity-centric trust model, data-centric trust model, and combined trust model.

At first, the entity-centric trust model aims to evaluate the trustworthiness of vehicles. The main difficulty is to accurately evaluate the trust of vehicles in the case of few connections between vehicles. This phenomenon is called Data Sparsity. Secondly, the data-centric trust model focuses on evaluating the trustworthiness of the messages. The main disadvantage is that the data-centric trust model can only temporarily evaluate the trust of the received messages, and can not form the trust relationship between vehicles. Finally, the combined trust model combines the entity-centric trust model and data-centric trust model.

In addition, with the development of blockchain, there are more studies consider applying blockchain to trust management in VANETs. However, there are still some defects in applying blockchain to VANETs. The blockchain-based model needs consensus from multiple parties, and its efficiency is much lower than that of the centralized model. Therefore, in the scenario of rapid response to emergencies, it is still a long way to go to apply blockchain technology to trust management in VANETs.

13.2 Basic Concepts of Trust in VANETs

As a subjective and fuzzy evaluation standard, trust can be used in the interaction process of vehicles. By evaluating the interaction process, it can provide a good basis and complete the decision, to improve the safety of vehicle interaction.

13.2.1 Characteristics of Trust

13.2.1.1 Asymmetry

Trust has obvious subjectivity, and the trust of both sides of the vehicle is not symmetrical. Even if they experience the same event together, the feedback or evaluation given by both sides are not the same, or even far from the same (for example, A trusts B, but it doesn't mean B trusts A).

13.2.1.2 Timeliness

The trust relationship is dynamic and time decay is an important factor in trust relationships. Even if vehicle A trusts vehicle B at a certain time, if they lack interaction and contact for some time, the trust value will decrease with time.

13.2.1.3 Limited Transitivity

Trust is transitive. Suppose that vehicle A trusts vehicle B and vehicle B trusts vehicle C, then vehicle A trusts vehicle C. However, vehicle A doesn't fully trust vehicle C. The trust degree of vehicle A to vehicle C is affected by the trust degree of vehicle A to vehicle B. The transitivity of trust decreases with the number of vehicles passing between them.

13.2.1.4 Content Relevance

Trust is for specific content, and trust between entities will not spread. For example, vehicle A trusts the identity provided by vehicle B, but this does not mean that vehicle A will trust vehicle B about its link-state, data integrity, and other information.

13.2.1.5 Regional

Vehicle A's trust in vehicle B is only applicable to a certain extent and does not unconditionally trust any judgment of vehicle B. This restriction is related to the degree of the event. (For example, vehicle A only believes in financial transactions of vehicle B with a certain amount or less.)

13.2.2 Common Malicious Attacks in Trust Evaluation for VANETs

The attack methods in VANETs are usually divided into usability, privacy, authenticity, integrity, and non-repudiation [3] (Fig. 13.1).

13.2.2.1 Usability

Availability ensures that networks and applications remain running even in the presence of error or malicious conditions.

(1) DoS attack: internal or external attackers execute DoS by interfering with the communication channel or covering the resources in the network. Attackers can

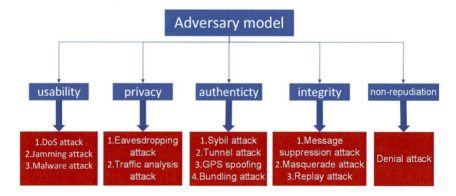

Fig. 13.1 Adversary model

be distributed, known as distributed denial of service (DDoS). The main goal is to prevent authorized vehicles from accessing services.

(2) Jamming attack: the attacker uses the strong signal of equivalent frequency to jam the communication channel.

(3) Malware attack: when the malware is installed into OBU and RSU, the attacker can penetrate VANETs and destroy the normal function.

(4) Broadcast jamming attack: internal attackers can broadcast false warning information, which will mask the correct security information to authorized vehicles.

(5) Black hole and gray hole attacks: black hole attacks and gray hole attacks will lose packets when forwarding packets in the network.

(6) Greedy behavior attack: malicious vehicles abuse media access control protocol to increase bandwidth at the expense of other vehicles.

(7) Spam attack: the attacker injects a large amount of spam into the virtual network system, which occupies the system bandwidth and causes system collision.

13.2.2.2 Privacy

It ensures that only the designated receiver can access the data, while the external vehicles cannot understand the privacy information belonging to each entity. Password solutions are usually used to solve privacy problems.

(1) Eavesdropping attack: the purpose of eavesdropping is to extract confidential information from protected data.

(2) Traffic analysis attack: attackers monitor message transmission, analyze its frequency and duration, and collect confidential information.

13.2.2.3 Authenticity

Authentication is a mechanism to protect the virtual network from malicious entity attacks. It is considered as the first line of defense against various attacks in VANETs.

(1) Sybil attack: a Sybil vehicle can forge many false identities to disrupt the normal operation of the virtual network, and then wrongly inform other vehicles that there is a traffic jam, forcing them to change their routes and make the road clear.

(2) Tunnel attack: tunnel attack is similar to wormhole attack. Attackers connect two remote parts of the network through tunnels or additional communication channels. Therefore, long-distance vehicles can communicate as neighbors.

(3) GPS spoofing: an attacker can generate a fake GPS signal stronger than the original signal from a trusted satellite, spoofing the vehicle into thinking it's available in a different location.

(4) Bundling attack: in the cooperative authentication scheme, selfish vehicles can make use of others' authentication contributions without self-authentication, which will bring serious threats to cooperative message authentication.

13.2.2.4 Integrity

Integrity ensures that the contents of messages are not modified during transmission, thus preventing unauthorized data creation, destruction, or modification.

(1) Message suppression attack: the attacker changes part of the transmission message to cause unauthorized effect.
(2) Masquerade attack: masquerade attackers can use the stolen password as valid user input VANETs to broadcast false messages.
(3) Replay attack: the attacker may continuously inject the previously received beacons and messages into the network, which will make it difficult for the traffic management department to identify them in an emergency.

13.2.2.5 Non-Repudiation

The non-repudiation ensures that the sender and receiver of the message cannot deny the transmission and reception of the message in case of dispute.

(1) Denial attack: attackers can deny the fact that they send or receive critical messages in case of disputes.

13.2.3 Trust Value Calculation

Trust value calculation modeling is the most basic problem of trust mechanism. The traditional calculation method considers single factors and the calculation formula is simple. With the development of the Internet, the context factors that affect trust increase, the types of data become larger, the value of data is sparse, and the noise increases. All these are important problems that affect the calculation of trust value. Trust value calculation is divided into two parts: direct trust calculation and indirect trust calculation [4].

Direct trust refers to the subjective trust generated by the subject, in the process of direct interaction between the subject and the object. Indirect trust refers to the infrequent or no direct interaction between the evaluated subject and the object, and the trust relationship between them is based on the trust relationship of other vehicles. The indirect trust is the evaluation result of other nodes on the object nodes. The relationship between direct trust and indirect trust is shown in Fig. 13.2.

13.2.3.1 Direct Trust Calculation

It analyzes and calculates according to the direct interaction information between the entity object and the target object. The key point is feature selection. The selected feature should not only summarize the characteristics of trust but also not cause the problem of overfitting. Currently, global iterative algorithms such as weighted

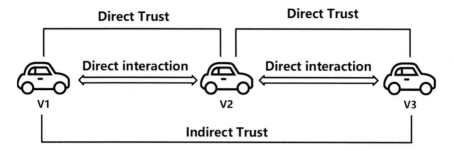

Fig. 13.2 Direct/Indirect trust relationship

average manner and statistical analysis manner can be selected. In addition, local aggregation algorithms such as analytic hierarchy process and context aggregation can also be selected.

13.2.3.2 Indirect Trust Calculation

Indirect trust is to solve the problem of the lack of direct interaction information between the entity object and the target object. It needs to find their common friends or indirect information for analysis and calculation. It is useful to help new users or users with less interaction accurately calculate the trust value. There are many difficulties in the calculation of indirect trust.

The first is the selection of similar vehicles. In addition to finding the intersection set, it also involves multi-layer trust transfer and similarity calculation. Many scholars consider using artificial intelligence, path recommendation, and other algorithms for optimization.

Secondly, the credibility of recommendation can be determined by collaborative filtering or unsupervised learning. In addition, there is another difficulty in trust value calculation: weight optimization. We should determine the relationship between multiple factors through time changes, increased interaction, context changes, etc. Dynamically adjusting the weight factor is one of the important means to ensure the practicability and reliability of the algorithm. In the current research, probability and statistics, fuzzy decision, gray correlation, and so on are commonly used dynamic weight calculation methods.

13.2.4 Trust Management

13.2.4.1 Dynamic Incentive Mechanism of Trust

The trust value changes dynamically as the interaction accumulates, and this change is affected by: the frequency of the interaction, the change of the interaction context, the decay of time, and so on. Therefore, how to establish a dynamic incentive and punishment mechanism is the focus of trust research.

The traditional dynamic incentive based on the results is to accumulate the results through the statistical analysis of the success and failure times of the results. However, this method does not consider the influence of context factors or propose dynamic incentives and penalties for evaluation. If there is a collision attack, there is no specific punishment for the cheater, which will seriously affect the accuracy of trust. To avoid the influence of malicious behaviors and ensure the stability of the trust mechanism, it is necessary to enrich and improve the incentive and punishment mechanism.

13.2.5 Trust Management Process

This section introduces the overall process of trust management against the background that vehicles seek cooperation from other trusted vehicles. The specific process is shown in Fig. 13.3. The following is a detailed description.

Assuming that the vehicle A wants to cooperate with the vehicle B, vehicle A will send a trust request to obtain vehicle B's global trust value to the RSU. If RSU locally stores B's global trust value, RSU will send B's global trust value to A. Otherwise, RSU will send a query request to the TA. The global trust distribution module is responsible for querying the TA's global trust database and returning the queried trust value. After obtaining the global trust value of vehicle B, vehicle A judges whether to trust vehicle B based on the set trust decision rules. If vehicle B is trustworthy, vehicle A interacts with vehicle B and updates the interaction records. Otherwise, vehicle A reselects a vehicle with higher global trust value. After the connection is interrupted, vehicle A performs direct trust calculation based on vehicle B's behavior and historical interaction records. The communication between A and B is completed.

Whenever a vehicle passes through the RSU, the local trust information of the vehicle will be uploaded to the RSU, and the RSU will aggregate it and upload it to the TA. Then, the TA periodically performs global trust calculation and stores the global trust value in the TA's global trust database.

13 Decentralized Trust Management System for VANETs

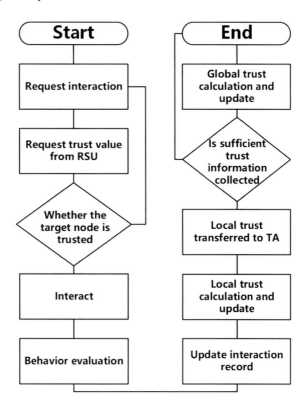

Fig. 13.3 Trust management process

13.2.6 Trust Management Architecture

Generally speaking, the trust management architecture is divided into three layers: vehicle trust management layer, local trust management layer, and global trust management layer [5].

13.2.6.1 Vehicle Trust Management Layer

Generally, the on-board trust management layer is deployed on the vehicle, including the trust factor collection module and local trust evaluation module [6].

The trust factor collection module is responsible for collecting evidence or trust factors and collecting different trust factors according to different scenarios, such as packet transmission rate, vehicle speed, quality of service, etc.

The local trust evaluation module is the most important module of vehicle trust management. This module uses the historical interaction records provided by the trust factor collection module to calculate the local trust value between vehicles. Common evidence aggregation methods include Bayesian inference [7], Dempster-Shafer evidence theory [8], fuzzy logic [9], and so on.

The trust decision module takes different behavior decisions according to different scenarios, such as selecting the vehicle with the highest local trust value for interaction.

13.2.6.2 Local Trust Management Layer

The local trust management layer is usually deployed on the RSU and is mainly responsible for data collection and dissemination. The corresponding includes two modules: local trust collection module and global trust dissemination module.

The local trust collection module is responsible for collecting the local trust evaluation information of the vehicle. When the vehicle enters the communication range of the corresponding RSU, it will actively upload its local trust evaluation of other vehicles to the RSU.

The function of the global trust module is to transfer the global trust information updated by the TA to the vehicle in need after the TA updates the global trust of the vehicle.

13.2.6.3 Global Trust Management Layer

The global trust management layer is usually managed by TA or the cloud, which is responsible for global trust calculation and data sharing. Therefore, it includes two modules: the global trust calculation module and the global trust data sharing module.

The global trust calculation module is the most important in this layer. It is responsible for the calculation of global trust. The calculation of global trust reflects the credibility of vehicles in the whole trust management system, rather than only the supervisor of a vehicle.

The global trust data sharing module periodically distributes the latest local global trust information to vehicles to ensure high accuracy and high real-time trust evaluation.

13.3 Classification of Trust Management in VANETs and Current Solutions

Trust mechanism model: in all interactive environments, the energy or resources of the vehicles participating in the interaction will be limited, while technologies with high energy consumption such as intrusion detection and password encryption and decryption technology are not suitable for the environment composed of resource limited terminal devices. The trust mechanism is to evaluate the trust of the evaluated vehicles by using the interaction records between vehicles in the system, which

13 Decentralized Trust Management System for VANETs

Table 13.1 Classification of trust mechanisms

Classification method	Concrete content
Entity-based trust model	The entity-based trust model aims to evaluate the credibility of vehicles
Data-based trust model	The data-based trust model focuses on evaluating the reliability of the received data
Combined trust model	The combined trust model is equivalent to the combination of the above two types of trust models
Blockchain-based trust model	The trust mechanism uses blockchain technology for trust data storage, and generally selects semi trusted RSU to act as miner node

provides a low-energy and highly reliable method for users' applications. It can also continuously update the trust of vehicles in the system, abandon the total malicious or unreliable and unstable vehicles in the system environment, and screen out more trusted vehicles for data interaction.

The trust model is a defense mechanism for evaluating the trust degree between equipment vehicles in the system environment. In the trust mechanism, the comprehensive evaluation values of directly interacting vehicles and other recommended vehicles are used to evaluate the trust degree of object vehicles.

With the popularity of the 5G network and the more and more extensive application of the distributed system, the computing mode in the era of Internet of things tends to develop in a distributed direction. The classification of the trust mechanism is shown in Table 13.1.

13.3.1 Entity-Based Trust Model

The entity-based trust model aims to evaluate the credibility of vehicles. Usually, a trust evaluation system can be constructed by using the direct trust or indirect recommendation trust between vehicles, to calculate the trust value of each vehicle, and detect untrusted or malicious vehicles.

The entity-based trust model is common. Tan et al. [10] used graph theory algorithm to evaluate the routing reliability of nodes based on two direct trust factors: packet transmission rate and average delay time. Xiao et al. [7] calculated local trust value between vehicles based on historical interaction information, then built a more stable trust link graph, and used the PageRank-based algorithm to calculate the global trust value of nodes. This helps vehicles to select trusted other vehicles for cooperation.

The entity-based trust model maintains the trust relationship between vehicles. The main difficulty lies in how to collect sufficient trust evidence to evaluate the trust of vehicles. Especially, due to the lack of sufficient interaction information, it is difficult to evaluate the trustworthiness of vehicles when vehicles have just joined VANETs.

13.3.2 Data-Based Trust Model

The data-based trust model focuses on evaluating the reliability of the received data. To accurately verify the reliability of the received data, the model needs to collect messages from a variety of information sources, such as neighbors and RSUs, and filter out those untrusted message data.

Huang et al. [11] studied and designed a novel voting mechanism based on the distance between the vehicle and the reported event. The closer the vehicle is to the event, the higher the voting weight. The vehicle will decide whether to believe the received event message according to the final voting result. Rawat et al. [12] used received signal strength and vehicle geographic location, to evaluate the trust level of received messages. Then, the study combined Bayesian estimation evaluation algorithm with determining distance calculation method to help identify malicious messages.

The main disadvantage of the data-based trust model is that the trust relationship of vehicles can never be formed, and only a short-term trust can be established for the received data. Because the data trust is based on each event, it is necessary to build the trust relationship for each event again and again, and the previous trust data has not been used. At the same time, when the number of received messages is not enough, it is difficult to judge the accuracy of messages.

13.3.3 Combined Trust Model

The combined trust model is equivalent to the combination of the above two types of trust models. It not only evaluates the trust degree of vehicles but also calculates the reliability of message data. Usually, the two kinds of trust are interrelated, that is, the trust value of vehicles affects the reliability of data to a certain extent, and the trust value of data, in turn, reflects the reliability of vehicles.

The attack-resistant trust management scheme proposed by Li et al. [13] estimates the reliability of vehicles and messages. Firstly, the scheme collects message data from multiple vehicles, uses Dempster-Shafer evidence theory to evaluate the reliability of messages, and then uses a collaborative filtering algorithm to calculate the reliability of vehicles based on functional trust factor and recommendation trust factor. However, the accuracy of the model is poor when there are few vehicles or sparse data. Farhan et al. [14] proposed a trust model to resist man-in-the-middle attacks. Firstly, the trust value of the sender is evaluated by performing a multi-dimensional entity-centered trust evaluation. After the receiver verifies the credibility of the sender, the credibility of the data content is evaluated by a combination of direct and indirect methods, although the scheme is more accurate compared with other models, the recall rate is greatly improved, but the evaluation scheme is more complex and the system overhead is greatly increased.

The combined trust model inherits the advantages and disadvantages of the entity-based and data-based trust models. The trust model not only needs to build a complete trust relationship between vehicles, but also can evaluate the reliability of each message data. Compared with the first two trust models, the trust evaluation process is more complex and the system overhead will be large. At the same time, the model also has problems such as sparse data.

13.3.4 Blockchain-Based Trust Model

In recent years, blockchain technology has become more and more mature and widely used. At present, some papers consider applying blockchain technology to VANETs trust management.

Xiao et al. [2] proposed a blockchain-based distributed trust management system (DTMS). The system adopted a consensus-based trust evaluation model and a blockchain-based trust storage system, providing a transparent evaluation process and irreversible trust storage. Yang [15] and others also proposed a decentralized VANETs trust management system based on blockchain technology. The innovation of this paper is that the authors proposed a consensus mechanism combining Proof of Work and Proof of Stack, which can ensure that all RSUs jointly maintain a newer, more reliable, and consistent trust blockchain. Ezedin et al. [16] proposed a trust evaluation method based on blockchain technology, which can resist false message attacks. This method combines proof of work and proof of stack in the process of miner election, to reduce the consumption of computing resources and network delay. However, this scheme is still not suitable for some delay-sensitive scenarios of Internet of vehicles.

The trust management system based on blockchain has considerable application prospects. Due to the decentralized characteristics of blockchain technology, the trust model based on blockchain can completely decentralize trust evaluation and management, greatly reducing the management and operation cost. At the same time, due to the tamper-proof characteristics of blockchain technology, it makes the blockchain-based trust model more robust and more vulnerable to attack than centralized trust management. However, blockchain technology is not omnipotent. Since the blockchain system requires multi-party vehicles to reach a consensus, its accounting efficiency is lower than that of the centralized management system; At the same time, the trust value of each vehicle should change dynamically, and the trust evaluation needs to be real-time. However, due to the non-repudiation of the blockchain, the trust value can not tamper once it is connected to the chain, which is contrary to the real-time performance of the trust value. Therefore, it is still a long way to go to apply blockchain technology to VANETs trust management [17].

13.4 Existing Problems and Future Research Focus

The existing trust model for VANETs may also have the following problems.

(1) Lack of trust model considering multiple application scenarios. At present, the trust model for VANETs only considers a single application scenario, namely communication message trust, and lacks consideration of vehicle behavior trust. However, the data trust and vehicle trust are two aspects that affect, complement, and confirm each other. To evaluate the trustworthiness of vehicles accurately and improve the accuracy of the trust model, it is important to consider multiple application scenarios.

(2) Lack of trust calculation method supporting dynamic update. Due to the timeliness of trust, the trust model needs to dynamically update the trust value. However, the dynamic update of trust value is ignored by existing algorithms (e.g., analytical method, matrix operation method, fuzzy logic method, probability method). It is necessary to study trust calculation and its algorithm to improve the real-time performance of trust model for VANETs.

(3) The dynamic adaptability of the trust decision-making mechanism is insufficient. Vehicles can obtain their benefits or cause losses to others through strategic behavior changes. Existing trust decision-making mechanisms generally have insufficient dynamic adaptability, and are difficult to deal with such dynamic strategic behavior changes of vehicles. The fraud or swing behavior of vehicle changing strategy is not detected and punished effectively.

Existing related technologies have important reference value for the study of Internet of vehicles trust model. However, few studies have considered the diversification of communication scenarios for VANETs. It is necessary to design and implement the trust management from a diversity perspective. In addition, it is necessary to focus on the application of blockchain and 5G-related technologies in trust management. It is the ultimate goal to build a trust model that can evaluate the trustworthiness of vehicles and messages accurately.

13.5 Conclusion

This chapter first introduced the basic concepts of trust in VANETs, including the basic attributes of trust, different types of attacks often suffered in VANETs, common methods of trust calculation, and the widely used trust management model. Then, the common trust models were classified and introduced in detail. The classification methods were: data-centric, entity-centric, hybrid, and fully distributed based on blockchain. Finally, the chapter put forward the current challenges and future research hotspots of VANETs trust management, which may give readers some enlightening thinking.

References

1. R. Hussain, J. Lee, S. Zeadally, Trust in VANET: a survey of current solutions and future research opportunities. IEEE Trans. Intell. Transp. Syst. **22**(5), 2553–2571 (2021). https://doi.org/10.1109/TITS.2020.2973715
2. X. Chen, J. Ding, Z. Lu, A decentralized trust management system for intelligent transportation environments. in *IEEE Transactions on Intelligent Transportation Systems*. https://doi.org/10.1109/TITS.2020.3013279
3. Z. Lu, G. Qu, Z. Liu, A survey on recent advances in vehicular network security, trust, and privacy. IEEE Trans. Intell. Transp. Syst. **20**(2), 760–776 (2019). https://doi.org/10.1109/TITS.2018.2818888
4. C. Zhang, R. Lu, X. Lin, P. Ho, X. Shen, An efficient identity-based batch verification scheme for vehicular sensor networks. *IEEE INFOCOM 2008—The 27th Conference on Computer Communications*, pp. 246–250 (2008). https://doi.org/10.1109/INFOCOM.2008.58
5. Chen, F. Bao, M. Chang, J. Cho, Dynamic trust management for delay tolerant networks and its application to secure routing. IEEE Trans. Parallel Distrib. Syst. **25**(5), 1200–1210 (2014). https://doi.org/10.1109/TPDS.2013.116
6. C.A. Kerrache et al., TACASHI: trust-aware communication architecture for social internet of vehicles. IEEE Internet Things J. **6**(4), 5870–5877 (2019). https://doi.org/10.1109/JIOT.2018.2880332
7. Y. Xiao, Y. Liu, BayesTrust and VehicleRank: constructing an implicit web of trust in VANET. IEEE Trans. Veh. Technol. **68**(3), 2850–2864 (2019). https://doi.org/10.1109/TVT.2019.2894056
8. H. Xia, S. Zhang, Y. Li, Z. Pan, X. Peng, X. Cheng, An attack-resistant trust inference model for securing routing in vehicular Ad Hoc networks. IEEE Trans. Veh. Technol. **68**(7), 7108–7120 (2019). https://doi.org/10.1109/TVT.2019.2919681
9. T.J. Willink, Possibility-based trust for mobile wireless networks. IEEE Trans. Mobile Comput. **19**(8), 1896–1909 (2020). https://doi.org/10.1109/TMC.2019.2911945
10. S. Tan, X. Li, Q. Dong, A trust management system for securing data plane of Ad-Hoc networks. IEEE Trans. Veh. Technol. **65**(9), 7579–7592 (2016). https://doi.org/10.1109/TVT.2015.2495325
11. H. Zhen et al., A social network approach to trust management in VANETs. Peer-to-Peer Netwo. Appl. **7**(3), 229–242 (2014)
12. Rawat, Danda B., et al. Trust on the security of wireless vehicular Ad-hoc networking. Ad Hoc Sensor Wirel. Netw. **24**(3–4), 283–305 (2014)
13. W. Li, H. Song, ART: an attack-resistant trust management scheme for securing vehicular Ad Hoc networks. IEEE Trans. Intell. Transp. Syst. **17**(4), 960–969 (2016). https://doi.org/10.1109/TITS.2015.2494017
14. F. Ahmad, F. Kurugollu, A. Adnane, R. Hussain, F. Hussain, MARINE: man-in-the-middle attack resistant trust model in connected vehicles. IEEE Internet Things J. **7**(4), 3310–3322 (2020). https://doi.org/10.1109/JIOT.2020.2967568
15. Z. Yang, K. Yang, L. Lei, K. Zheng, V.C.M. Leung, Blockchain-based decentralized trust management in vehicular networks. IEEE Internet Things J. **6**(2), 1495–1505 (2019). https://doi.org/10.1109/JIOT.2018.2836144
16. E. Barka et al., Towards a trusted unmanned aerial system using blockchain for the protection of critical infrastructure. Trans. Emerg. Telecommun. Technol., 2(2019)
17. B. Luo, X. Li, J. Weng, J. Guo, J. Ma, Blockchain enabled trust-based location privacy protection scheme in VANET. IEEE Trans. Veh. Technol. **69**(2), 2034–2048 (2020). https://doi.org/10.1109/TVT.2019.2957744
18. Y. Wang et al., CATrust: context-aware trust management for service-oriented Ad Hoc networks. IEEE Trans. Serv. Comput. **11**(6), 908–921 (2018). https://doi.org/10.1109/TSC.2016.2587259

Chapter 14
Intrusion Detection System for Connected Automobiles Security

Abdul Majid Jamil⑩, Jianhua Zhou, Di Wang, Hassan Jalil Hadi, and Yue Cao

Abstract Modern automobiles are controlled via networked controls. The majority of networks were built with minimal regard to security. Researchers expect to see a variety of attacks on the system recently, prompting them to share their data. This chapter discusses the weaknesses of the Controller Area Network (CAN) within the in-automobile communication protocol and several possible attacks that may be used against it. In addition, we present recent security detection schemes that have been offered in the current level of research to counteract the threats. The primary purpose of this study is to showcase an integrated technique known as an intrusion detection system (IDS). It has been a significant instrument in protecting information and networks systems for decades. Therefore, we have presented a detailed literature review examining existing IDS. For the investigation of IDS, we considered the following aspects: approaches used for detection, strategies used for deployment, attack mechanisms, and technical issues and challenges. We classify and compare current IDS approaches based on criteria such as real-time limitations, hardware types, CAN Bus behavior changes, attack mitigation types, and software/hardware used to validate these systems. Similarly, other scholars will be encouraged to pursue IDS research on the CAN bus system as a result of the current study.

Keywords Intrusion detection system · In-automobile Security · IDS Attacks and Defense CAN bus Architecture and its Security · Anomaly Based Detection

---

A. M. Jamil (✉) · J. Zhou · D. Wang · H. J. Hadi · Y. Cao
School of Cyber Science and Engineering, Wuhan University, Wuhan, China
e-mail: 2021272210001@whu.edu.cn

J. Zhou
e-mail: 390992612@qq.com

D. Wang
e-mail: 1065889068@qq.com

Y. Cao
e-mail: yue.cao@whu.edu.cn

© The Author(s), under exclusive license to Springer Nature Singapore Pte Ltd. 2023
Y. Cao et al. (eds.), *Automated and Electric Vehicle: Design, Informatics and Sustainability*, Recent Advancements in Connected Autonomous Vehicle Technologies 3, https://doi.org/10.1007/978-981-19-5751-2_14

Abbreviations

V2I	Automobile-to-Infrastructures
V2V	Automobile-to-Automobile
ECUs	Electronic Control Units
CAN	Controller Area Network
LIN	Local Area Network
LAN	Local Interconnect Network
FOTA	Firmware Update over the Air
MOST	Media Oriented Systems Transport
SOF	Start of Frame
ACK	Acknowledgement
AVB	Audio Video Bridging
TSN	Time-Sensitive Networking
EOF	End of Frame
IDS	Intrusion Detection System
NIDS	Network Based Intrusion Detection System
HIDS	Host Based Intrusion Detection System
DDoS	Distributed Denial-of-Service
DT	Decision Trees
NN	Neural Networks
SVM	Support Vector Machines
HMM	Hidden Markov Models
ML-IDS	Machine Learning Intrusion Detection System

14.1 Introduction

In the last decade, car manufacturers have transformed the shape and functionalities of modern automobiles by rapidly adopting many novel technologies. Automation and connectivity with the outside world are advanced characteristics. With increasing safety and enabling automobile collaboration through Automobile-to-Infrastructures (V2I), Automobile-to-Automobile (V2V), intelligent automobiles have been offering safer driving experiences. Electronic Control Units (ECUs), sensors, actuators, and communication devices are the core components of modern automobiles.

The sensors can aid the automobile's perception of surroundings, automate several ECU to perform corresponding functions of driving tasks. The ECU is grouped into sub-networks based on their functions (steering angle control, powertrain, infotainment, etc.). The automobile network is structured by interconnecting the sub-networks via gateways. The Controller Area Network (CAN) is a network protocol for automobile communication and helps to connect the ECU. Particularly, CAN is suitable for intra-automobile networking due to its lightweight operation cost and reliable fault tolerance. However, it has been proven that CAN is vulnerable to cyber-attacks, which means that many modern automobiles might be susceptible to new threats due to their access to external networks. The lack of data authentication and broadcasting nature in the CAN bus protocol are two critical reasons for vulnerability.

Since the CAN bus protocol originally does not include an authentication mechanism to verify the origin of the communications, attackers can manipulate and relay their injected data easily to influence the automobile system. The author has demonstrated that a wide range of attack vectors (Bluetooth, MP3 players, and communication devices) have already been utilized to remotely control the automobile illegitimately [1]. As a result, automobile security has become anxiety among the general public, a critical problem in the automobile industry, and a source of concern among the general public.

Protecting automobiles from cyber-attacks is of importance since cars have traditionally been constructed without considering complete security needs, with the assumption that automobiles are self-contained and have no communication infrastructure [2]. Because of time sensitivity, resource constraints, and complexity of cars, typical proactive security countermeasures like encryption, access control, and the like may not be appropriate. Despite the amount of research on connected automobile security in literature, the author [3] believes that a systematic assessment of state-of-the-art automobile IDS can be a helpful contribution, in providing a new and critical overview of the most current studies on this issue. However, new security vulnerabilities arise as the number of informed transactions between CAN buses and buses, given an expanded interface with the outside world. These flaws in the CAN bus might threaten the driving safety of automobiles with potential impact on those in proximity.

The Contribution of this chapter discusses the weaknesses of the Controller Area Network (CAN) within the automobile communication protocol, and several possible attacks that may be used against it. In addition, an in-depth review of existing literature [4–9] related to CAN security is provided. Vulnerabilities in CAN bus protocol and how they threaten security are explored in this paper. Identification of potential challenges and open issues linked with CAN. In addition, this chapter presents recent security detection schemes that have been offered in the current level of research to counteract the threats. Finally, this paper addresses the current approaches used for intrusion detection, strategies used for deployment, and attack mechanisms.

14.1.1 External Communication In-Automobile

As automobiles become more revolutionary, connectivity has become a requirement. To control multiple operations of the car remotely, various communication protocols are used. Manufacturers are attempting to provide consumers with more user-friendly options for controlling other components of their automobiles remotely, in addition to the typical radio-controlled door unlocking function. Here, cellular connectivity (GSM, 3G, and 4G) and WiFi (IEEE 802.11) are two examples of increasingly

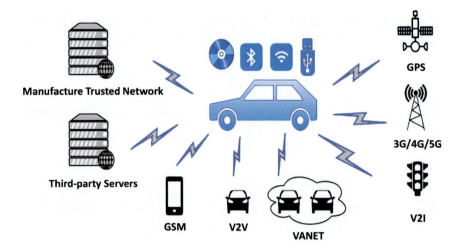

Fig. 14.1 Internal communication system of In-automobile

common options. These communication technologies generally control typical car functions, such as turning on the engine or air conditioning [10] (consumers may purchase all of these technologies and others as stock options).

Additionally, diagnostics data is required from the perspective of the manufacturer, for pushing remote Firmware Update Over the Air (FOTA), to ensure the most outstanding aftermarket experience for customers. Manufacturers may also legally collect periodic reports from automobiles [11], where VPNs provided by the manufacturer or third-party services are typically used for these conversations.

To demonstrate how communication technologies influence modern automobiles in terms of design and driving experience, Fig. 14.1 shows a breakdown of the communication technologies that impact cars.

14.1.2 Internal Communication In-Automobile

One or more Electronic Control Units (ECU) control almost every function in a modern car. A compact embedded computer system with real-time computation, time limits, and minimal power consumption is known as an ECU. The primary responsibility of the ECU is to collaborate to share sensor data via messages. The ECUs are divided into five different groups:

- Powertrain
- Comfort-train
- Safety
- Infotainment
- Telematics [12]

Powertrain

The powertrain is divided into two critical resource controls as transmission control and engine management. The brake system, which assists the driver in preventing wheel sliding, and the distribution of electrical braking force, which maintains the stability of the automobile, are examples of designs in this category. These ECUs are critical in terms of safety, as their failure could result in the driver's losing control of the car.

Comfort-train

Anti-lock braking systems, adaptive cruise control, airbags, tire pressure monitoring, and collision avoidance systems are all included in the automobile safety category. These ECUs are also critical in safety, as their failure to work could result in driver casualties.

Safety

ECUs that provide driver assistance, such as electronic suspension, thermal management, and parking aid, are included in the safety category. Because the category's goal is to aid the driver, the ECU's functionality failures may not immediately impact the driver's safety.

Infotainment

The infotainment category includes audio and video support systems for automobiles. Digital streaming television, audio streams, and TFT displays are such systems. This category also includes systems that accept data from external sources, such as traffic and weather information systems.

Telematics

Systems that merge telecommunications and informatics fall under the telematics category. This type of system is utilized to offer automobile networked software applications. This category also includes ECUs that provide mobile connectivity, such as GPRS.

A network of buses and ECUs in an automobile is known as the brake system. It assists the driver in preventing wheel sliding and the brake force distribution, maintaining automobile stability. It is an example in this category of the automobile network.

Multiple communication protocols, such as CAN, Local Interconnect Network (LIN), Flex Ray, MOST, and Ethernet connect the ECU [13]. To coordinate and collaborate on various activities, they transmit and receive data across many automobiles' Local Area Network (LAN) domains. If the sender and recipient are on different LANs, special ECU known as gateways are used to transfer data from one domain to another. The construction of an internal automobile network is depicted in Fig. 14.2. The Control Area Network (CAN) is the most popular bus technology in current cars, it provides efficient, fast, reliable, and cost-effective communication between electronic control units (ECUs).

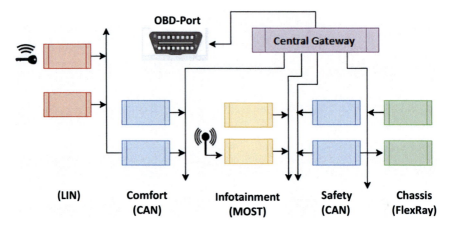

Fig. 14.2 Internal communication system in-automobile

In addition, it allows the connected ECU to send and receive data across the bus utilizing only one channel. Another serial bus intercommunication system similar to CAN is the LIN, while the latter is less expensive and has been widely applied for applications without real-time requirements [13]. The FlexRay is superior to CAN for high-speed communication with the time-critical requirement, such as those services in backbone system, drive-by-wire, and brake-by-wire. Unlike CAN, FlexRay supports two-channel communication and a variety of topologies, including star, ring et al., whereas it is expensive in terms of price [13]. FlexRay allows for high-speed communication with synchronization and eliminates buffering, lag, and jitter.

Another high-speed bus communication standard for multimedia networks in automobiles with a ring topology is Media Oriented Systems Transport (MOST). MOST is compatible with large bandwidth and carries video and audio data [13]. Audio Video Bridging (Ethernet AVB) and its Time-Sensitive Networking (Ethernet TSN IEEE 802.1) are increasingly utilized to carry audio and video data in current automobile multimedia fields.

The goal of this paper is to provide an understanding of CAN behavior within automobile networks and its vulnerabilities, to demonstrate a variety of potential attacks on the CAN bus system, security mechanism.

14.1.3 Controller Area Network (CAN)

Understanding the underlying system's operation may aid in fully understanding the vulnerability of the system. As a result, this subsection provides an overview of CAN protocol and strategies for CAN bus attacks defense.

14.1.3.1 Frame Structure of CAN Data

The CAN bus includes a variety of electric components across the automobile network, such as sensors devices, ECU, micro-controllers, and actuators. The CAN operates by broadcasting packets, which means all components connected to the CAN bus will receive broadcasted packets [14]. Here, the frame is a structure in a CAN packet that carries a sequence of Controller Area Network (CAN) data bytes via a network. Each transmitted CAN frame arbitration ID field determines the precedence of packets, where a lower ID bit value indicates a higher packet priority. With this, it can prevent traffic collisions within the automobile, by scheduling the dissemination order of CAN packets.

The four types of CAN data frames are data frames, error frames, overloaded frames, and remote frames [15]. Arbitration identification, acknowledgment, data, and a few additional elements make up the typical structure of CAN packets.

The CAN packet ID field usually has an 11-bit value format, while expanding to 29 bits. On the other hand, the Controller Area Network (CAN) data field carries data from 0 to 64 bits. The semantics of the data field and CAN ID are secret and closely guarded by automobile manufacturers. Figure 14.3 shows the basic architecture of the CAN data frame for both the ordinary and advanced arbitration identification versions.

The following is a list of all the fields found in the CAN data frame:

a. Start of Frame (SOF): It indicates the starting bit of Controller Area Network (CAN) data, with a dominant bit and alerts.
b. Arbitration: It is a method of resolving disputes. As previously stated, the ID field is made up of 11 bits and can be expanded towards a format of (29) bits.
c. Control: Also called a monitor field, it tells the receiver whether all of the intended packets were properly received.
d. Data: It contains content information that CAN nodes need to understand and perform, with the size ranging from (0 to 8) bytes.
e. CRC: The cyclic redundancy code, often known as the defense field, is a (15) bit flaw detecting method that verifies the authenticity of packets.
f. (ACK): It's also called the confirmation field. This field ensures that the CAN packets are received accurately by the recipient nodes. The transmitter will be informed instantly by the receiver if an error is detected during the transmission process, and the data packets will be sent again.
g. End of Frame (EOF): A recessive bit's flag marks the end of the CAN frame.

The CAN data frame is the sole frame utilized to send CAN packet data. As a result, the CAN bus transmits packets over data frames to connect with other (ECU) nodes. It transforms into a CAN data frame when the RTR (remote transmission request) is a dominant bit.

Fig. 14.3 Frame structure of CAN data

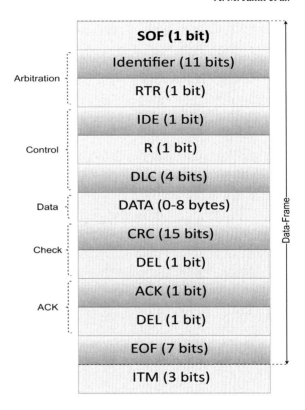

14.1.3.2 CAN Vulnerabilities

In the CAN bus system, there are multiple significant vulnerabilities. As previously indicated, CAN packets do not contain any transmitter or receiver address information. The arbitration identifier field sends all CAN packets to ECU nodes [16]. As a result, because the source of the packets is not provided, the receiving nodes are unable to determine whether the packets received are intended for them.

As a result of this fact, the receiving node was unable to establish whether the packet received was genuine or not. Furthermore, no message authentication mechanism exists in each ECU to secure packets sent between nodes. As a result, attackers might spoof and broadcast bogus CAN packets through hacked ECUs [17]. Because of the flaws mentioned above, a CAN bus system is vulnerable and unable to determine which nodes are responsible for the assaults.

14.2 Related Work on Automobiles Security

This section reviews the current literature on automobile security, particularly CAN bus technologies. Automobile security research has just lately become available. Certain active studies and initiatives have been published to secure the in-automobile CAN bus system. Message authentication is one of the most widely used security techniques in cryptographic software [5]. The literature that presented this technique takes inspiration from internet security to resolve issues of CAN network security. This method ensures the validity of the CAN data frame sent between the two end nodes. The CAN data frame's maximum length is limited to 8 bytes despite this. As a result, there is a finite amount of space available to implement this procedure. Several countermeasures have been used to overcome the problem, such as overwriting a MAC across multiple CAN frames, adding 8 bytes of CBC-MAC, using different CRC values, and authenticating the message through an out-of-band channel [6]. However, depending solely on this strategy cannot guarantee perfect security to prevent a high degree of danger, particularly a denial-of-service (DoS) assault. Furthermore, this solution only safeguards a limited set of automobile components and demands a significant update to all ECUs, which is impractical.

Verendel et al. [7] presented a honeypot security mechanism installed at the wireless gateway and operated as a decoy in replicating the in-automobile network. Any information gathered about the attacks was analyzed to improve future iterations of the system. One of the difficulties with honeypot deployment is that it must be as realistic as possible so that the attacker does not understand he is not attempting to access the "real" network. To safeguard vehicular communication gateways, Wolf presented a firewall concept architecture. Only authorized controllers could deliver the valid message across the CAN bus network, thanks to firewall signatures (a set of customizable restrictions). However, the authors suggested that a virtual security device such as a firewall would be insufficient to safeguard the car network because most modern automobiles include internal diagnostic ports that enable access to the complete automobile system.

Several companies have been working on various aspects of in-automobile network system risks. For example, Arilou Cyber Security offers a groundbreaking parallel intrusion prevention system (PIPS), which determines the source of each CAN packet on the bus. It also uses an electronic signature to distinguish between the signals sent by the various ECUs. In this method, tainted May packet transmission can be avoided.

VeCure is an automobile-based applied security technique that addresses the CAN Bus message authentication issue, according to Wand and Sanjay et al. [9]. The VeCure approach has the following properties. It is designed to work with today's connected automobile technologies. It could also be used as a foundation for building trust. It is recognized as a groundbreaking message authentication method that can compute offline, reducing the time and online message processing cost. VeCure uses CAN message authentication to isolate fake messages injected from a compromised or targeted ECU and OBD-II port.

14.3 Typical Attacks in CAN

The various types of attacks outlined in [8, 18–23] are considered in this research, based on the latest findings within the CAN network. Once attackers get access to the automobile from either inside or outside, they can launch sniffing attacks, fuzzing attacks, DoS attacks, and replay attacks on the CAN Bus network. The following are some of the mechanisms used to launch these attacks (Fig. 14.4).

14.3.1 Sniffing Attack on CAN Bus

Even without authentication measures, encryption, or broadcast transmission, sniffing the data on the CAN Bus is possible [24]. Using off-the-shelf OBD2 sniffers like the CAN board [25], it is possible to read and analyze data on the bus, as well as edit and originate associated messages [24]. To prevent CAN frames from being exposed, this attack can be addressed by encrypting them. Due to the passive nature of sniffing communications, this exploit is difficult to detect. The next stage is to reverse engineer CAN communications to target individual automobile components with them. This is an important step, and most manufacturers publish their CAN message requirements [26].

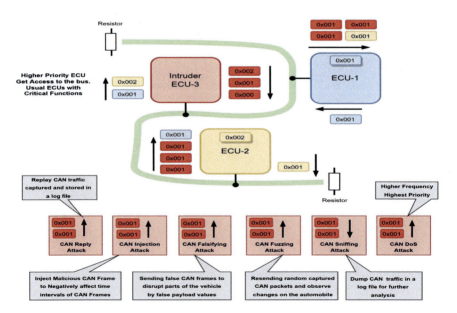

Fig. 14.4 Attacks on CAN bus network in automobiles

14.3.2 Fuzzing Attack on CAN Bus

ECUs accept and respond to CAN communications because the CAN Bus protocol lack of authentication and lacks of data integrity checks. This attack checks the bus and observes changes on the automobile's instrument panel using random CAN data frames. The impacts of CAN frames on ECUs are investigated in this attack, which includes monitoring changes in automobile speed while injecting CAN frames. It usually occur after intercepted CAN packets have been sniffed and analyzed. It can alternatively be created via a black-box method, in which the CAN ID and payload values are generated at random with no knowledge of the actual CAN ID. It entails randomly delivering recording the outcomes and CAN frames. To eliminate data analysis and to only receive CAN frames from legitimate ECUs, encryption, and authentication are required.

14.3.3 Falsifying Attack on CAN Bus

This exploit injects fake values into the payload of a CAN message. For example, the attacker might inject fake parameter values into an automobile. This type of modification attack is used to supply bogus data payload to interrupt car functions when the CAN ID is known. This is due to the CAN Bus protocol's inability to provide data integrity and authentication. To prevent this attack, CAN Bus should include authentication to authenticate the source of the data before acting on it. This form of assault typically contains only a little quantity of data, making it difficult to detect and monitor. To detect this attack, a system might consider checking CAN ID and data payload consistency in a time interval.

14.3.4 Injection Attack on CAN Bus

By data injecting into a CAN Bus [27], it is feasible to deliver messages at an unusual rate. This attack aims to alter the size and frequency of CAN frames on the bus, as well as the order in which valid CAN frames and data payloads are sequenced. Because CAN Bus does not provide authentication to check whether the sender is real, this attack would inject aberrant CAN traffic onto the bus, which would target the automobile speed. Any node can connect to the bus because there is no encryption. After that, the bus data can be processed to recover the arbitration and data fields, as well as generate messages to imitate activities [19]. This could result in the creation of bogus events that cause sections of the automobile to act how the attacker desires. Authentication and integrity methods can be used to prevent this attack. The broadcast frequency of CAN IDs may increase as a result of the attack, which can be detected by abnormal broadcast behavior (Table 14.1).

Table 14.1 Attacks on CAN bus network in automobiles

Authors	Attacks initiated	Gateways	Attackers positions	Result of attacks	Testbed
Eisenbarth et al. [28]	Sensors	• Key fob keyless • Entry point	External	Start Engine Lock or Unlock Door	Real Car
Koscher et al. [29]	Interfaces Infotainment	• USB • OBD-2 • CD Player	Internal Internal Internal	Injection in CAN Bus Full control	Real Car
Hoppe et al. [23]	Interfaces	• OBD-2	Internal	Airbag and warning light, Control Vehicle Glasses	CANoe Simulator, Parts of Car
Checkoway and McCoy [30]	Interfaces Infotainment Telematics	• Radio • CD Player • Cellular • Bluetooth • OBD-2	External Internal External External Internal	Get CAN bus access Disable vehicle parts	Real Car
Rouf et al. [31]	Sensors	• TPSM	External	Signal jamming and injecting false TPSM values	Real Car
Miller and Valasek [21]	Interfaces	• OBD-2	Internal	Control breaks and get CAN bus access	Real Car
Petit et al. [32]	Sensors	• Cameras • LiDAR	External External	Jamming signal	CAN Software LiDAR Hardware
Miller and Valasek [21]	Telematics	• WIFI	External	Stop engine by injecting CAN packet through unauthorized access	Jeep Cherokee
Nie et al. [33]	Telematics	• WIFI	External	Get CAN bus access	Tesla S model
Zorz [34]	Interfaces	• OBD-2 Cellular Dongle	External	Injection in CAN Bus	Real Car

14.3.5 CAN Bus DoS Attack

The CAN ID priority is utilized to access the medium with multi-access in both classic CAN and CAN FD [20]. The arbitration field is utilized by the nodes on the CAN Bus to determine which node can occupy the bus and transfer data and which message has priority. In this situation, a DoS attack can be launched by exploiting the priority arbitration method based on the CAN frame and sending too many highest priority frames, preventing other nodes from using the bus by high attribution id, like 0×000, to dominate the bus and make it busy by sending too much high priority frames, preventing other nodes from using the bus. [21]. It can also leverage an existing ECU's CAN message-id and, knowing its transmission rate, do a DoS by increasing the frequency time. If an ECU transmits a message every 200 ms, for example, the attacker can raise the frequency by injecting the identical message at a higher frequency, causing the sensor portion to be disrupted.

14.4 Defense of Connected Automobiles

An intrusion detection system (IDS) attracts much attention due to its ease of use and ability to identify attacks quickly. In general, an IDS monitor's network or host activity detects and raises an alert if any unusual occurrences occur in the system.

Depending on how it was set up, IDS might be passive or active. Active IDS intervenes to prevent the attack, whereas inactive IDS detects it. Sensors, a detection engine, and a reporting module in that sequence make up a typical IDS design. Sensors can be connected to the network or the end node directly (network and host IDS).

14.4.1 IDS in the Automobile Industry

Using an intrusion detection system (IDS) to detect malicious attempts is a common practice in-automobile networks. IDS based on signatures or anomalies are used [23]. The position of the IDS is also crucial: Based on host-IDS vs. network-IDS [35]. HIDS to detect attacks [36] might not be feasible or cost-effective for current automobile networks because it would require ECU change. As a result, adding a NIDS to the CAN Bus as an additional node, such as an OBD-2 dongle, is easier and more practical, and requires no CAN Bus changes [37]. An intrusion detection system (IDS) can be passive, in the sense that it just records intrusions, or active, in the sense that it actively prevents attacks. Even if nothing changes, ECUs inside the car send CAN messages at regular intervals [38]. As a result, IDSs are created using modifications of typical CAN traffic behavior. Another way to look for anomalies is by analyzing changes in each ECU's physical layer features, such as its voltage profile and signal.

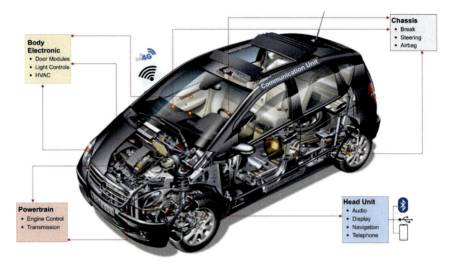

Fig. 14.5 Possible places for IDS installation in the controller area network system

14.4.2 Deployment Plan for IDS

IDS must be installed to each monitored activity and system in the controller area network from various sources. Authors recommended these areas appropriate for IDS deployment, based on studies to achieve this in the automobile context [39]. (A) CAN network, (B) ECUs, and (C) Central gateways are the three types of CAN networks are described in Fig. 14.6. It provides a comprehensive perspective of the system's internal activity. As a result, malicious scripts injected can be discovered during execution. On the other hand, Network-based IDS integrates to the central gateways and CAN network. It inspects and monitors the onboard automobile communication system to detect active attacks (Fig. 14.5).

However, some issues need to be considered when implementing an intrusion detection system into an in-automobile. Kleberger et al. [40] studied the effects of assaults on ECUs located in different positions in the automobile communication system in previous work. Because hacked networks and central gateways have access to and control packets, they suffer from a much higher risk than when ECU is hacked.

14.5 IDS Attacks Detection Approach

14.5.1 Signature Based IDS

This IDS is based on the detection of a pre-defined set of threat signatures. Despite having a low false-positive rate in the detection technique, when new threats emerge,

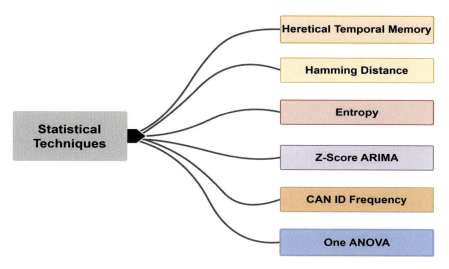

Fig. 14.6 Statistical based approach IDS

it must update its database signatures [22]. Signature-based IDSs must also maintain a massive database of known automobile network threats [41]. It might be difficult and time-consuming to extract attack signatures for a moving target in real-time.

14.5.2 Anomaly Based IDS

This strategy is implemented using statistical, signature-based, and ML techniques. It develops a learning algorithm that can detect odd traffic, identify new patterns, and predict assaults never seen before. Statistical techniques are used This IDS learns the system's regular behavior via conditional statistical association analysis, as shown in Fig. 14.7. A baseline pattern is produced as a threshold if departures from the norm are identified. CAN attribute like payload consistency and CAN ID frequency are used in statistical analysis in CAN Bus networks. CAN frames are sent by ECUs at specified intervals. These CAN message contain a unique CAN identity, as well as the frame time interval and frames numbers in each time unit [42]. Furthermore, the payloads inside CAN frames usually include consistent sequential values. Correlations between automobile parameters like speed and RPM signals are part of a larger strategy (statistical relationship between speed and RPM in normal operation). Finally, data field semantics, message transmission frequency, message ID, amount of packets received during a defined time frame, and message received sequence can all be used [43]. As a result, this IDS can detect tampered with or erroneous payload values, as well as CAN ID usage that is inconsistent.

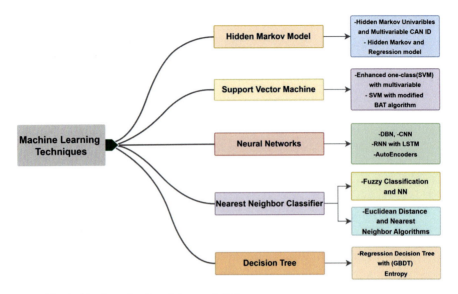

Fig. 14.7 Machine Learning-Based Approach IDS

14.5.2.1 Using Statistical Approaches

The following statistical approaches have been described in the literature, and each of these approaches is discussed in further detail further below.

Frequency of CAN ID: Feng and Ling [44] employ abnormalities in traffic frequency to detect DoS and injection attacks. ECU impersonated assaults and payload manipulation attacks, in which the attacker ECU produces authentic CAN ID messages, are difficult to detect utilizing their small-size attack technique.

CAN frames Intervals: Shin and Cho [45] created a clock-based IDS that fingerprints each ECU based on the interval between message exchanges. They use the least square cost function and a sequential analysis technique called the Cumulative Sum algorithm to find anomalies. As with the prior technique, low-size injected messages and impersonated disabled ECUs are difficult to detect. The testbed is made up of an Arduino UNO board and a Seed Studio CAN shield. Other approaches used an algorithm to evaluate and detect unusual time periods for certain CAN ID frame broadcasts, such as the one described in Reference [46]. They planned to concentrate on CAN injection attacks.

CAN Traffic Behavior across a Time Window: Based on the assessment of data flow abnormalities, the author of Reference [47] constructed an IDS. This method involves comparing current CAN traffic statistical values to past ones over a 1 s period. On the other hand, Small-volume malicious transmissions are impossible to identify precisely with this anomaly detection over time [48].

14 Intrusion Detection System for Connected Automobiles Security

One-way ANOVA Function: Reference [49] employed a one-way ANOVA method to statistically determine the pattern within a dataset and generated a set of normal patterns to find irregularities. They separated the CAN dataset into groups depending on automobile metrics like fuel consumption, engine parameters, gear ratio, and discover aberrant events.

Hamming Distance: Many researchers have looked at anomaly detection using Hamming distance methods. For example, in Ref. [41], the authors looked at the CAN payload and captured every bit in the data field. Attacks were identified if the computed Hamming distance function deviated significantly from the calculated Hamming distance function.

Absolute differences and quantized interval: For periodic CAN ID, the authors of Reference [50] used a quantized interval anomaly detection system and examined the absolute difference of the CAN payload values using a quantized interval anomaly detection system. Their solution was proven to be successful against message injection attacks (measures like False Positive/Negative and True Positive/Negative). Low-size injection assaults, on the other hand, were difficult to detect using their approach, according to them.

Cumulative Sum algorithm: The statistical cumulative sum approach was utilized by the authors in Reference [51] to create an anomaly detection system. This is a change detection strategy that uses sequential analysis.

In a Time Defined Window, Z-SCORE and ARIMA: When employing Z-SCORE and ARIMA in a set time window, Reference [52] used an average value for the numerous times a CAN ID was broadcast throughout a time interval. This was used to see how the broadcast intervals of CAN ID varied over time and over different periods of time. To compare the mean broadcast intervals of CAN ID, the author developed a Z-Score, ARIMA, and a supervised approach. Dropped packet attacks and CAN injections were identified.

Heretical Temporal Memory: Reference [53] employed a distributed IDS based on Heretical Temporal Memory, a technique frequently used in time-series analysis and forecasting (Table 14.2).

14.5.2.2 Machine Learning Approaches

IDS that employ Machine Learning (ML) to extract and learn normal vs. abnormal behavior, then create a model to identify and forecast assaults, could be a good fit. With its many features, ML-IDS is widely used to handle large amounts of CAN traffic data. It's convenient to be able to extract and pre-process raw CAN data. This is especially important because car companies rarely provide detailed specifications or guidance on how to interpret raw data features. Because raw CAN data must be tagged, the CAN attack must be identified, and the data must then be labeled and categorized, supervised machine learning techniques can be time-consuming.

Table 14.2 Statistical based approach IDS

Authors	Types	Layer	Payload/CAN ID	Method of detection	Detected attacks	Online dataset
Ling and Feng [44]	CAN ID Frequency	Data Link (Controller layer)	CAN behavior Identity	Malicious CAN IDs should be detected. Irregular CAN Frequency can be detected	• DoS • Injection	Not Available
Hsu et al. [49]	One-Way ANOVA	Datalink	Consistency of data payload across numerous CAN signals	Compare the average of comparable CAN frames based on a standard statistical observation, such as engine power and speed	• Manipulation Data Payload	Not Available
Marchetti et al. [54]	Anomaly detection based on entropy	Data Link	The frequency of CAN IDs has varied	For each group or class of CAN communications, provide independent variables for an entropy-based anomaly detector	• Manipulation Data Payload • Payload that has been tampered	Not Available
Stabili et al. [41]	Using Hamming distance to detect intrusions	Data Link	CAN message-ids with consecutive data payloads	In the consecutive data payloads of CAN message ID, compare the variations in Hamming distance values	• Spoofing • Injection	Not Available

(continued)

Table 14.2 (continued)

Authors	Types	Layer	Payload/CAN ID	Method of detection	Detected attacks	Online dataset
Marchetti and Stabili [27]	Identification of anomalies using the ID sequence	Datalink	CAN ID messages are sent in a specific order	Compare the CAN ID sequence to the knowledge gained from the real-time model	• Injection • Replay	Not Available
Tomlinson et al. [52]	Detecting anomalies in message timing in time windows to detect attacks	Datalink	In a specific time window, CAN ID broadcast means	Detection of an attack based on a set of rules	• DoS Attacks • Injections	Not Available
Tomlinson et al. [42]	Algorithm for time-series data. Z-Score and ARIMA	Datalink	Intervals of broadcast in a time window	Detection of an attack based on a set of rules	• Drop • Injection	Not Available
Lee et al. [48]	Distant frame and offset ratio IDS	Datalink	Intervals for CAN requests and responses using a remote frame	Changes in time intervals and the resulting shift in reaction to a remote frame	• Impersonation of an ECU • Injection	Available [55]
Wang et al. [56]	CAN ID-based entropy IDS	Datalink	Each CAN ID's entropy	Changes in each bit of the CAN ID should be detected	• Injection • Flooding	Not Available
Olufowobi et al. [51]	In a specific time window, use the Cumulative Sum procedure	Datalink	A sequence of CAN IDs	Sequence behavior of the CAN ID	• Attack of the Frame Fuzz • DoS Attack • Injection	Not Available

Unsupervised machine learning methods, on the other hand, do not require labeled datasets, and the algorithms can discover common patterns in data and use them to classify traffic and detect anomalous behavior.

Hidden Markov Models: This method detects anomalous behavior using time-series data. Narayanan et al. [57] built a model that could detect anomalies and trigger warnings using an IDS based on an HMM. They looked into using each ID individually as well as combining numerous automobile factors, such as automobile speed and RPM CAN ID messages. They tested their model using real-time data against rapid changes in speed and RPM by sending malicious messages to each parameter individually. They tested a variety of techniques by combining malicious speed and RPM messages.

Support Vector Machines: The authors of Reference [58] used an unsupervised ML technique to improve one-class Support Vector Machines (SVM) to operate with several factors to categorize CAN data. Their method learned normal behavior from unlabeled time-series data from an automobile and detected anomalies based on departures. They employed a training set of real automobiles with error-free logs in their technique. The researchers next utilized a noisy data model to discover mistakes and anomalies in the recorded data. The authors of Reference [59] employed one-class SVM and compared it to Random Forest and conventional One-class SVM (Using the True Positive Rate metric, this leads to improved detection accuracy).

Deep Learning/Neural Networks (NN): Deep neural networks were employed by the authors of Reference [60] to learn regular patterns from unstructured samples and recognize deviations as anomalies. To pre-process the data and find a regular trend, they employed an unsupervised Deep Believe Network. They used noise to imitate real automotive data in their test dataset to validate their method. They used a real-world automotive test bench and network experimentation software to simulate and create CAN frames [61]. Reference [62] employed a Recurrent Neural Network with Long Short-Term Memory to detect CAN Bus attacks.

Their solution works with raw CAN Bus data, removing the need for data reduction and abstraction during the pre-processing stage of the study. To detect patterns in CAN data without classifying it, the authors of Reference [63] used Generative Adversarial Nets. In terms of DoS, frame fuzzification, and spoofing attacks, they put their solution to the test. Convolutional Neural Networks were used by the authors of Reference [64] to build an IDS capable of recognizing successive patterns of vehicular traffic in order to DoS attempts and detect spoofing. Their approach implies that CAN traffic is transmitted straight into their model without any processing. They tested their method offline and admitted that putting it online in existing automobiles is difficult.

Decision trees: Approaches based on decision trees (DT) divide CAN data into two categories (normal, anomalous). To make judgments, decision tree techniques require a supervised labeled dataset during the training step. The authors of Reference [14] used a regression Decision Tree with a Gradient Boosting approach to building a classifier. To design the choosing approach, they evaluated the entropy of the CAN

14 Intrusion Detection System for Connected Automobiles Security

ID and the data payload time using entropy. Gradient Boosting is also a technique for obtaining the best DT model by training many trees. They tested their method with 750,000 messages using real recorded CAN data. They altered the test dataset by introducing anomalies in the form of random aberrant values.

Nearest Neighbor Classifier: To distinguish attacks targeting the CAN Bus, the authors employed fuzzy classification techniques based on Nearest Neighbor classifiers in Reference [65]. To validate their method, they used an online dataset containing CAN attacks. To identify CAN traffic, they employed the data payload actual data (8 bytes) as a feature. They put their method to the test using the dataset's DoS, frame injection, and frame fuzzification attacks. Using a neural network approach, they were able to reach a precision of 0.85 to 1. This detector, on the other hand, may miss minor injected fake messages and ECU impersonated attacks (Table 14.3).

14.5.3 Comparing Both IDS

An IDS is used to identify malicious CAN attacks, as previously indicated. Signature-based IDS has been demonstrated to identify attacks with a minimal number of false positives; however, a CAN Bus attack signature must be obtained. As a result, new CAN attacks may be difficult to detect. It's also difficult to extract attack signatures and CAN signals when a car is moving. Anomaly or behavior-based IDS has the benefit of being able to identify and forecast attacks based on the training and learning process, and it may be utilized without additional training in some circumstances. Raw data obtained directly from automobiles can help IDS based on machine learning. Automobiles generate large volumes of data that must be pre-processed before it can be used, therefore machine learning algorithms can handle a wide variety of factors. Unsupervised machine learning can uncover patterns and detect abnormalities in unlabeled raw data, which may be used to detect anomalies (Table 14.4).

14.6 Limitation with Intrusion Detection System

Limited resources, time constraints, traffic pattern behavior, unpredictably significant connection, sizes, weight, and cost are just a few of the difficulties that must be considered when developing suggested solutions for the CAN bus network system [73].

1. **Resources Limitation**: It refers to resources of all electronic control units (ECUs) in the car, such as memory, computing power, and data transmission speed.
2. **Time Constraints**: The transmission delay occurred through CAN bus protocol and queued delay occurred through buffer are critical. Meanwhile, the CAN

278   A. M. Jamil et al.

Table 14.3 ML-based approach IDS

Authors	Type	Threshold detection	Mechanism of detection	Detected attacks	Online dataset
Theissler [58]	SVM improved	Data payload and CAN ID CAN signals with many variants	Enhanced one-class Support Vector Machine (ESVM) deviation	• Error and Signal faults	Not Available
Narayanan et al. [57]	Hidden Markov Model	Univariate and Multivariant CAN signal e.g., RPM and speed	Deviation from the expected pattern of behavior	• Single-injection • Multiple injections	Not Available
Taylor et al. [47]	Long short-term memory RNN	Data payload behavior and CAN ID	The RRN model's devastation and the LSTM mechanism's observations	• Injection • DoS	Not Available
Kang and Kang [60]	Neural Network for Deep Belief with Probability	CAN ID and data payload behavior	Change away from the NN model	• Injection	Not Available
Chockalingam et al. [66]	O-SVM	Intervals and frequencies of CAN messages Anomaly IDS with a single class of support	Anomaly detection with a single SVM class	• Fuzzing	Not Available
Martinelli et al. [65]	Uncertainty in classification Classification by Nearest Neighbor	Data payload and CAN ID	Anomalies are detected by inspecting each byte of the data payload for characteristics	• DoS • Injection • Fuzzy	Available [67]
Levi et al. [68]	Regression Model Using the Hidden Markov Model	To establish a threshold for the log probabilities, we used an HMM and a regression model	Offline dataset learning and online dataset learning	• Noise Attack	Not Available

(continued)

14 Intrusion Detection System for Connected Automobiles Security

Table 14.3 (continued)

Authors	Type	Threshold detection	Mechanism of detection	Detected attacks	Online dataset
Tomlinson et al. [42]	Nearest neighbor and Euclidean distance algorithms	In the time window, the CAN ID frequency	Modifications to the CAN ID broadcast data payload	• Fuzzy	Not Available
Hanselmann et al. [69]	GBDT Entropy Regression	Changes in CAN ID and data payload entropy	CAN traffic entropy change	• Injection • DoS	Not Available
[55]	Machine Learning Hybrid-IDS	The payload sequence of CAN messages in the static check module. Online Algorithm LODA and RNN-based IDS OCSVM	Payload consistency and detection in the time window	• Injection • DoS	Not Available
Wang et al. [53]	HMS-based Multiple Anomaly IDS	Anomaly in data sequence based on HMS	From an online stream, multiple HMS-IDS for each CAN signal are learned	• Injection • DoS	Not Available
Avatefipour et al. [70]	O-SVM with a BAT algorithm that has been modified	Message-IDs that deviate from repeating patterns	SVM algorithm using only one class	• Injection • DoS	Not Available
Weber et al. [71]	Deep Conventional Neural Network	Detects message injection by learning sequential patterns of CAN Bus traffic	Conventional Neural Network	• Message Injection	Available [72]

Table 14.4 IDS based on statistical, ML are compared

Type of IDS	Layer	Data gathering	Prevention/detection	Implementation costs
Statistical IDS	Layer2	The behavior of the CAN ID Frame/Frame Flow	Yes/No	Data that is both medium and needed to be labeled
Machine Learning IDS	Layer2	The behavior of the CAN ID Frame/Frame Flow	Yes/No	Medium. Can work with both labeled and unlabeled data

packet received by in-automobile sensors must be regulated in real-time to enable actuators to complete their functions as rapidly as feasible. The real-time limitation must be considered while offering secure solutions.

3. **Traffic patterns**: The CAN traffic protocol for automobile communication systems is different from that IP traffic standard in Internet. For example, CAN packets are broadcasted for in-automobile communication. Furthermore, to do automobile firmware changes and wireless diagnostics, a temporary link must be established between the automobile and the infrastructure, requiring new communication models and traffic protocols. As a result, implementing an IP network solution is unfeasible.

4. **Unstable connections**: Automobiles may move into areas with no internet connections because they move regardless of speed. As a result, Connecting IDS to a third party is not always feasible.

5. **Size, price, and weight**: Intrusion detection system solutions are available in various sizes and weights they may require few or substantial modifications, all of which might impact the price. Removing a line-topology CAN bus network and replacing it with a different topology ethernet bus network, for example, would necessitate a substantial quantity of additional cabling and may be impractical [74].

14.7 Future and Research Direction

Future research might benefit from comparing the computing efficiency of different data preprocessing methods. The response system's implementation may necessitate further adjusting the design of a distinct IDS component. When the reaction mechanism detects something unexpected, it can activate the security mode right away, allowing the automobile to be parked safely [8]. Because it requires coordination from multiple components, this response system may be more challenging to implement than improving detection performance. As a result, the primary purpose of the IDS response system should be to install intelligent human-automobile interaction and a rapid action mechanism. The majority of the research findings focused on a single IDS module. However, this may not include a complete plan for meeting the security

14 Intrusion Detection System for Connected Automobiles Security

demands of the automobile communication system [8]. It should be supplemented with a more lightweight cryptography-based approach, such as message authentication. Finally, despite the rising interest in developing improved IDS techniques for the CAN bus system, only a few research have compared with those in other similar cases. It is critical to investigate and verify the key differences between the proposed technology and different approaches to achieve optimal performance of related procedures.

14.8 Conclusion

This book chapter explored characteristics of behavior and vulnerabilities in the CAN bus system, offering a review of the standard CAN packet structure applied to an automobile communication system. In addition, we explored several types of potential attack surfaces in the CAN bus according to the current literature, ranging from direct physical access to remote wireless access. We also reviewed recent security remedies that researchers have developed to tackle this vulnerability. We also introduce intrusion detection systems that provide comprehensive protection for traditional networks, and intrusion detection systems protect the CAN communication system in the automobile sector.

As a result, in a literature search specifically for in-automobile CAN bus systems, we investigated the state-of-the-art IDS to extend past work in offering various security countermeasures techniques. We reviewed different research papers from the literature that suggested different IDS approaches and tactics for detecting and mitigating threats. The selected articles range from 2010 to 2020 and are divided into four areas: IDS detection methodology, deployment tactics, attack techniques, and technical hurdles. According to our views, IDS research for the automobile sector is gaining popularity. Despite this, we believe that the issues presented in this study will inspire academics with further efforts to promote innovation in this area. It necessitates security researchers to conduct in-depth investigations into various attacking approaches and methods to produce accurate data for training and testing.

References

1. S.A. Alasadi, W.S. Bhaya, Review of data preprocessing techniques in data mining **12**(16), 4102–4107 (2017)
2. J.P. Amaral et al., Policy and network-based intrusion detection system for IPv6-enabled wireless sensor networks, in *2014 IEEE International Conference on Communications (ICC)*. IEEE (2014)
3. L. Apvrille et al., Secure automobile onboard electronics network architecture, in F*ISITA 2010 World Automobile Congress*, Budapest, Hungary (2010)
4. K.-T. Cho, K.G. Shin, Fingerprinting electronic control units for automobile intrusion detection, in *25th USENIX Security Symposium (USENIX Security 16)* (2016)

5. D. Erhan et al., Why does unsupervised pre-training help deep learning? in *Proceedings of the Thirteenth International Conference on Artificial Intelligence and Statistics. JMLR Workshop and Conference Proceedings* (2010)
6. B. Groza et al., Libra-can: a lightweight broadcast authentication protocol for controller area networks, in *International Conference on Cryptology and Network Security*. Springer (2012)
7. S. Halder, M. Conti, S.K. Das, COIDS: a clock offset based intrusion detection system for controller area networks, in *Proceedings of the 21st International Conference on Distributed Computing and Networking* (2020)
8. S. Woo, H.J. Jo, D.H. Lee, A practical wireless attack on the connected car and security protocol for in-automobile CAN **16**(2), 993–1006 (2014)
9. Q. Wang, S. Sawhney. VeCure: a practical security framework to protect the CAN bus of automobiles, in *2014 International Conference on the Internet of Things (IoT)*. IEEE (2014)
10. A. Avalappampatty Sivasamy, B. Sundan, A dynamic intrusion detection system based on multivariate Hotelling's T2 statistics approach for network environments (2015)
11. V.S. Barletta et al., A Kohonen SOM architecture for intrusion detection on in-automobile communication networks **10**(15), 5062 (2020)
12. M.-J. Kang, J.-W. Kang, A novel intrusion detection method using deep neural network for in-automobile network security, in *2016 IEEE 83rd Vehicular Technology Conference (VTC Spring)*. IEEE (2016)
13. M. Müter, A. Groll, F.C. Freiling, A structured approach to anomaly detection for in-automobile networks, in *2010 Sixth International Conference on Information Assurance and Security*. IEEE (2010)
14. V.S. Barletta et al., Intrusion detection for in-automobile communication networks: an unsupervised Kohonen som approach **12**(7), 119 (2020)
15. H. Boyes, A. Luck, A security-minded approach to automobile automation, road infrastructure technology, and connectivity (2015)
16. I. Butun et al., A survey of intrusion detection systems in wireless sensor networks **16**(1), 266–282 (2013)
17. P. Carsten et al., In-automobile networks: attacks, vulnerabilities, and proposed solutions, in *Proceedings of the 10th Annual Cyber and Information Security Research Conference* (2015)
18. R.I. Davis et al., Controller area network (can) schedulability analysis with fifo queues, in *2011 23rd Euromicro Conference on Real-Time Systems*. IEEE (2011)
19. Roderick Currie. 2017. Hacking the CAN bus: basic manipulation of a modern automobile through CAN bus reverse engineering. SANS Institute (2017). Retrieved August 2020 from https://www.sans.org/reading-room/whitepapers/threats/hacking-bus-basic-manipulation-modern-automobile-through-bus-reverse-engineering-37825
20. T. Huang, J. Zhou, Y. Wang, A. Cheng, On the security of in-automobile hybrid network: status and challenges. In Lecture Notes in Computer Science (including subseries Lecture Notes in Artificial Intelligence and Lecture Notes in Bioinformatics), Vol. 10701. Springer, Cham, 621–637 (2017)
21. C. Miller. C. Valasek, Remote exploitation of an unaltered passenger automobile. Defcon 23, 2015(2015), 1–91. Retrieved August 2020 https://www.academia.edu/download/53311546/Remote_Car_Hacking.pdf
22. A.S. Syed Navaz, V. Sangeetha, C. Prabhadevi, A.S. Syed Navaz, V. Sangeetha, C. Prabhadevi, Entropy based anomaly detection system to prevent DDoS attacks in cloud. Int. J. Comput. Appl. **62**(15), 42–47 (2013)
23. T. Hoppe, S. Kiltz, J. Dittmann, Applying intrusion detection to automobile it—early insights and remaining challenges. J. Inf. Assur. Secur. **4**, 226–235 (2009)
24. Roderick Currie. 2015. Information security reading room developments in car hacking. Retrieved August 2020 from https://www.sans.org/reading-room/whitepapers/ICS/developments-car-hacking-36607
25. Netronics Ltd. 2020. CANdo—CAN Bus Analyser. Retrieved August 2020 from http://www.cananalyser.co.uk/index.html

26. T.U. Kang, H.M. Song, S. Jeong, H.K. Kim, Automated reverse engineering and attack for CAN using OBD-II, in *Proceedings of the IEEE Vehicular Technology Conference*, pp. 1–7 (2018)
27. Mi. Marchetti, D. Stabili, Anomaly detection of CAN bus messages through analysis of ID sequences, in *Proceedings of the IEEE Intelligent Automobiles Symposium*, pp. 1577–1583 (2017)
28. T. Eisenbarth, T. Kasper, A. Moradi, C. Paar, M. Salmasizadeh, M.T. Manzuri, Shalmani., On the power of power analysis in the real world: a complete break of the KeeLoq code hopping scheme, in *Lecture Notes in Computer Science (including subseries Lecture Notes in Artificial Intelligence and Lecture Notes in Bioinformatics)*, vol. 5157, (Springer, Berlin, 2008), pp.203–220
29. K. Koscher, A. Czeskis, F. Roesner, S. Patel, T. Kohno, S. Checkoway, D. Mccoy, B. Kantor, D. Anderson, H. Shacham, S. Savage, H. Snachám, S. Savage, Experimental security analysis of a modern automobile, in *Proceedings of the IEEE Symposium on Security and Privacy*, pp. 447–462 (2010)
30. S. Checkoway, D. McCoy, Comprehensive experimental analyses of automobile attack surfaces, in *Proceedings of the 20th USENIX Conference on Security* (2011), 6 (2011)
31. I. Rouf, R. Miller, H. Mustafa, T. Taylor, S. Oh, We. Xu, M. Gruteser, W. Trappe, I. Seskar, Security and privacy vulnerabilities of in-car wireless networks: a tire pressure monitoring system case study, in *Proceedings of the 19th USENIX Security Symposium*, pp. 323–338 (2010)
32. J. Petit, B. Stottelaar, M. Feiri, F. Kargl, Remote Attacks on Automated Automobiles Sensors: Experiments on Camera and LiDAR. Blackhat.com, pp. 1–13 (2015). Retrieved January 2021 from https://www.blackhat.com/docs/eu-15/materials/eu-15-Petit-Self-Driving-And-Connected-Cars-Fooling-SensorsAnd-Tracking-Drivers-wp1.pdf
33. S. Nie, L. Liu, Y. Du, Free-fall: hacking tesla from wireless to can bus. Defcon, 1–16 (2017). Retrieved on August 2020 from https://www.blackhat.com/docs/us-17/thursday/us-17-Nie-Free-Fall-Hacking-TeslaFrom-Wireless-To-CAN-Bus-wp.pdf
34. Z. Zorz, Backdooring connected cars for covert remote control—Help Net Security (2018). Retrieved August 2020 from https://www.helpnetsecurity.com/2018/03/05/backdooring-connected-cars/
35. W. Wu, R. Li, G. Xie, J. An, Y. Bai, J. Zhou, K. Li, A survey of intrusion detection for in-automobile networks. IEEE Trans. Intell. Transport. Syst. **21**(3), 919–933 (2019)
36. U.E. Larson, D.K. Nilsson, E. Jonsson, An approach to specification-based attack detection for in-automobile networks, in Proceedings of the IEEE Intelligent Automobiles Symposium, pp. 220–225 (2008)
37. C. Young, H. Olufowobi, G. Bloom, J. Zambreno, Automobile intrusion detection based on constant CAN message frequencies across automobile driving modes, in *Proceedings of the ACM Workshop on Automobile Cybersecurity*, pp. 9–14 (2019)
38. C. Miller, C. Valasek. OG Dynamite Edition (2016). Retrieved August 2020 from http://illmatics.com/can%20message%20injection.pdf
39. H. Ji et al., Comparative performance evaluation of intrusion detection methods for in-automobile networks **6**, 37523–37532 (2018)
40. K. Koscher et al., Experimental security analysis of a modern automobile, in *2010 IEEE Symposium on Security and Privacy*. 2010. IEEE Computer Society.
41. D. Stabili, M. Marchetti, M. Colajanni, Detecting attacks to internal automobile networks through Hamming distance, in *Proceedings of the 2017 AEIT International Annual Conference: Infrastructures for Energy and ICT: Opportunities for Fostering Innovation (AEIT'17)*, pp. 1–6 (2017)
42. A. Tomlinson, J. Bryans, S.A. Shaikh, Towards viable intrusion detection methods for the automobile controller area network, in *Proceedings of the Computer Science in Cars Conference (CSCS'18)* (2018)
43. H. Ji, Y. Wang, H. Qin, Y. Wang, H. Li, Comparative performance evaluation of intrusion detection methods for In-Automobile networks. IEEE Access **6**, 37523–37532 (2018)

44. C. Ling, D. Feng, Analgorithm for detection of malicious messages on CAN buses, in *Proceedings of the 2012 National Conference on Information Technology and Computer Science (CITCS'12)* (2012)
45. K.-T. Cho, K.G. Shin, Fingerprinting electronic control units for automobile intrusion detection, in *25th Usenix Security Symposium (Usenix Security'16)*, pp. 911–927 (2016)
46. H.M. Song, H.R. Kim, H.K. Kim, Intrusion detection system based on the analysis of time intervals of CAN messages for in-automobile network, in *Proceedings of the International Conference on Information Networking*, pp. 63–68 (2016)
47. A. Taylor, N. Japkowicz, S. Leblanc, Frequency-based anomaly detection for the automobile CAN bus, in *Proceedings of the 2015 World Congress on Industrial Control Systems Security (WCICSS'15)*. Infonomics Society, pp. 45–49 (2015)
48. H. Lee, S.H. Jeong, H.K. Kim, OTIDS: A novel intrusion detection system for inautomobile network by using remote frame, in *Proceedings of the 2017 15th Annual Conference on Privacy, Security and Trust (PST'17)*, pp. 57–66 (2018)
49. C.H. Hsu, F. Xia, X. Liu, S. Wang, Internet of automobiles—safe and intelligent mobility, in *Proceedings of the2nd International Conferenceon Internet of Automobiles (IOV'15)*. Lecture Notes in Computer Science (including subseries Lecture Notes in Artificial Intelligence and Lecture Notes in Bioinformatics) 9502 (2015), pp. 89–97 (2015)
50. T. Koyama, T. Shibahara, K. Hasegawa, Y. Okano, M. Tanaka, Y. Oshima, Anomaly detection for mixed transmission CAN messages using quantized intervals and absolute difference of payloads, in *Proceedings of the ACM Workshop on Automobile Cybersecurity, co-located with CODASPY 2019 (AutoSec'19)* (2019), pp. 19–24 (2019)
51. H. Olufowobi, U. Ezeobi, E. Muhati, G. Robinson, C. Young, J. Zambreno, G. Bloom, Anomaly detection approach using adaptive cumulative sum algorithm for controller area network, in *Proceedings of the ACM Workshop on Automobile Cybersecurity, co-located with CODASPY 2019 (AutoSec'19)*. ACM, pp. 25–30 (2019)
52. A. Tomlinson, J. Bryans, S.A. Shaikh, H.K. Kalutarage, Detection of automobile CAN cyberattacks by identifying packet timing anomalies in time windows, in *Proceedings of the 48th Annual IEEE/IFIP International Conference on Dependable Systems and Networks Workshops (DSN-W'18)*, pp. 231–238 (2018)
53. C. Wang, Z. Zhao, L. Gong, L. Zhu, Z. Liu, X. Cheng, A distributed anomaly detection system for in-automobile network using HTM. IEEE Access **6**(2018), 9091–9098 (2018)
54. M. Marchetti, D. Stabili, A. Guido, M. Colajanni, Evaluation of anomaly detection for in-automobile networks through information-theoretic algorithms, in *Proceedings of the 2016 IEEE 2nd International Forum on Research and Technologies for Society and Industry Leveraging a Better Tomorrow (RTSI'16)*, pp. 1–6 (2016)
55. Hacking and Countermeasure Research Lab. 2020. CAN-intrusion-dataset (OTIDS)—Hacking and Countermeasure Research Lab. Retrieved August 2020 from http://ocslab.hksecurity.net/Dataset
56. Q. Wang, Z. Lu, G. Qu, An entropy analysis-based intrusion detection system for controller area network in automobiles. Int. Syst. Chip Conf., 174–179 (2019)
57. S.N. Narayanan, S. Mittal, A. Joshi, OBD_SecureAlert: an anomaly detection system for automobiles, in *Proceedings of the 2016 IEEE International Conference on Smart Computing (SMARTCOMP'16)*. IEEE, 1–6 (2016)
58. A. Theissler, Anomaly detection in recordings from in-automobile networks, in *First International Workshop on Big Data Applications and Principles (BIGDAP'14)* (2014)
59. T. Dagan, A. Wool, Parrot, a software-only anti-spoofing defense system for the CAN bus, in *Proceedings of the 14th Embedded Security in Cars (ESCAR'16)*, 10 (2016)
60. M.J. Kang, J.W. Kang, Intrusion detection system using deep neural network for in-automobile network security. PLoS One **11**(6), 1–17 (2016)
61. P.N. Borazjani, C.E. Everett, D. Mccoy, OCTANE: an extensible open source car security testbed, in *Proceedings of the Embedded Security in Cars Conference (ESCAR'14)*, pp. 1–10 (2014)

14 Intrusion Detection System for Connected Automobiles Security

62. A. Taylor, S. Leblanc, N. Japkowicz, Anomaly detection in automobile control network data with long short-term memory networks, in *Proceedings of the 3rd IEEE International Conference on Data Science and Advanced Analytics (DSAA'16)*, pp. 130–139 (2016)
63. E. Seo, H.M. Song, H.K. Kim, GIDS: GAN based intrusion detection system for the in-automobile network, in *Proceedings of the 2018 16th Annual Conference on Privacy, Security and Trust (PST'18)* (2018)
64. H.M. Song, J. Woo, H.K. Kim, In-automobile network intrusion detection using deep convolutional neural network. Vehic. Commun. **21**(2020), 100198 (2020)
65. F.o Martinelli, F. Mercaldo, V. Nardone, A. Santone, Car hacking identification through fuzzy logic algorithms, in *Proceedings of the IEEE International Conference on Fuzzy Systems* (2017)
66. V. Chockalingam, I. Larson, D. Lin, S. Nofzinger, Detecting Attacks on the CAN Protocol With Machine Learning (2017). Retrieved August 2020 from http://www-personal.umich.edu/valli/assets/files/CAN_AD.pdf
67. HCRL. 2019. Hacking and Countermeasure Research Lab. Retrieved August 2020 from https://sites.google.com/a/hksecurity.net/ocslab/Datasets/car-hacking-dataset
68. M. Levi, Y. Allouche, A. Kontorovich, Advanced analytics for connected car cybersecurity, in *Proceedings of the IEEE Vehicular Technology Conference*, pp. 1–7 (2018)
69. M. Hanselmann, T. Strauss, K. Dormann, H. Ulmer, CANet: an unsupervised intrusion detection system for high dimensional CAN bus data (2020). . Retrieved from https://arxiv.org/abs/1906.02492
70. O. Avatefipour, A.S. Al-Sumaiti, A.M. El-Sherbeeny, E.M. Awwad, M.A. Elmeligy, M.A. Mohamed, H. Malik, An intelligent secured framework for cyberattack detection in electric automobiles' can bus using machine learning. IEEE Access **7**(2019), 127580–127592 (2019)
71. M. Weber, S. Klug, E. Sax, B. Zimmer, M. Weber, S. Klug, E. Sax, B. Zimmer, E. Hybrid, A. Detection, M. Weber, S. Klug, E. Sax, B. Zimmer, Embedded hybrid anomaly detection for automobile CAN communication, in *Proceedings of the Embedded Real Time Software and Systems (ERTS2'18)* (2018)
72. Hacking and Countermeasure Research Lab. 2020. Car-Hacking Dataset—Hacking and Countermeasure Research Lab. Retrieved August 2020 from http://ocslab.hksecurit.net/Datasets/CAN-intrusion-dataset
73. M. Marchetti, D. Stabili, Anomaly detection of CAN bus messages through analysis of ID sequences, in *2017 IEEE Intelligent Automobiles Symposium (IV)*. IEEE (2017)
74. M. Marchetti et al., Evaluation of anomaly detection for in-automobile networks through information-theoretic algorithms, in *2016 IEEE 2nd International Forum on Research and Technologies for Society and Industry Leveraging a Better Tomorrow (RTSI)*. IEEE (2016)

Printed in the United States
by Baker & Taylor Publisher Services